Numerical Analysis of Partial Differential Equations Using Maple and MATLAB

Fundamentals *of* Algorithms

Editor-in-Chief: Nicholas J. Higham, University of Manchester

The SIAM series on Fundamentals of Algorithms is a collection of short user-oriented books on state-of-the-art numerical methods. Written by experts, the books provide readers with sufficient knowledge to choose an appropriate method for an application and to understand the method's strengths and limitations. The books cover a range of topics drawn from numerical analysis and scientific computing. The intended audiences are researchers and practitioners using the methods and upper level undergraduates in mathematics, engineering, and computational science.

Books in this series not only provide the mathematical background for a method or class of methods used in solving a specific problem but also explain how the method can be developed into an algorithm and translated into software. The books describe the range of applicability of a method and give guidance on troubleshooting solvers and interpreting results. The theory is presented at a level accessible to the practitioner. MATLAB® software is the preferred language for codes presented since it can be used across a wide variety of platforms and is an excellent environment for prototyping, testing, and problem solving.

The series is intended to provide guides to numerical algorithms that are readily accessible, contain practical advice not easily found elsewhere, and include understandable codes that implement the algorithms.

Series Volumes

Aurentz, J. L., Mach, T., Robol, L., Vandebril, R., and Watkins, D. S., *Core-Chasing Algorithms for the Eigenvalue Problem*
Gander, M. J. and Kwok, F., *Numerical Analysis of Partial Differential Equations Using Maple and MATLAB*
Asch, M., Bocquet M., and Nodet, M., *Data Assimilation: Methods, Algorithms, and Applications*
Birgin, E. G. and Martínez, J. M., *Practical Augmented Lagrangian Methods for Constrained Optimization*
Bini, D. A., Iannazzo, B., and Meini, B., *Numerical Solution of Algebraic Riccati Equations*
Escalante, R. and Raydan, M., *Alternating Projection Methods*
Hansen, P. C., *Discrete Inverse Problems: Insight and Algorithms*
Modersitzki, J., *FAIR: Flexible Algorithms for Image Registration*
Chan, R. H.-F. and Jin, X.-Q., *An Introduction to Iterative Toeplitz Solvers*
Eldén, L., *Matrix Methods in Data Mining and Pattern Recognition*
Hansen, P. C., Nagy, J. G., and O'Leary, D. P., *Deblurring Images: Matrices, Spectra, and Filtering*
Davis, T. A., *Direct Methods for Sparse Linear Systems*
Kelley, C. T., *Solving Nonlinear Equations with Newton's Method*

Martin J. Gander
University of Geneva
Geneva, Switzerland

Felix Kwok
Hong Kong Baptist University
Kowloon Tong, Hong Kong

Numerical Analysis of Partial Differential Equations Using Maple and MATLAB

10 9 8 7 6 5 4 3 2 1

Publications Director	Kivmars H. Bowling
Executive Editor	Elizabeth Greenspan
Developmental Editor	Gina Rinelli Harris
Managing Editor	Kelly Thomas
Production Editor	Lisa Briggeman
Copy Editor	Bruce R. Owens
Production Manager	Donna Witzleben
Production Coordinator	Cally A. Shrader
Compositor	Cheryl Hufnagle
Graphic Designer	Doug Smock

Library of Congress Cataloging-in-Publication Data

Names: Gander, Martin J., author. | Kwok, Felix (Writer on scientific computing), author.

Title: Numerical analysis of partial differential equations using Maple and MATLAB / Martin J. Gander (University of Geneva), Felix Kwok (Hong Kong Baptist University).

Description: Philadelphia : Society for Industrial and Applied Mathematics, [2018] | Series: Fundamentals of algorithms ; 12 | Includes bibliographical references and index.

Identifiers: LCCN 2018023072 (print) | LCCN 2018029193 (ebook) | ISBN 9781611975314 | ISBN 9781611975307 | ISBN 9781611975307(print)

Subjects: LCSH: Differential equations, Partial--Numerical solutions--Data processing. | MATLAB. | Maple (Computer file)

Classification: LCC QA377 (ebook) | LCC QA377 .G233 2018 (print) | DDC 515/.353028553--dc23

LC record available at https://lccn.loc.gov/2018023072

Dedicated to our parents
Edith and Maurice Gander
Cecilia Ho and Pui Yan Kwok

Contents

Preface **ix**

1 Introduction **1**
1.1 Notation . . . 2
1.2 ODEs . . . 6
1.3 PDEs . . . 15
1.4 The Heat Equation . . . 17
1.5 The Advection-Reaction-Diffusion Equation . . . 25
1.6 The Wave Equation . . . 28
1.7 Maxwell's Equations . . . 31
1.8 Navier–Stokes Equations . . . 33
1.9 Elliptic Problems . . . 34
1.10 Problems . . . 38

2 The Finite Difference Method **45**
2.1 Finite Differences for the Two-Dimensional Poisson Equation . . . 46
2.2 Convergence Analysis . . . 51
2.3 More Accurate Approximations . . . 55
2.4 More General Boundary Conditions . . . 56
2.5 More General Differential Operators . . . 57
2.6 More General, Nonrectangular Domains . . . 62
2.7 Room Temperature Simulation Using Finite Differences . . . 63
2.8 Concluding Remarks . . . 66
2.9 Problems . . . 67

3 The Finite Volume Method **69**
3.1 Finite Volumes for a General Two-Dimensional Diffusion Equation 70
3.2 Boundary Conditions . . . 71
3.3 Relation between Finite Volumes and Finite Differences . . . 74
3.4 Finite Volume Methods Are Not Consistent . . . 76
3.5 Convergence Analysis . . . 80
3.6 Concluding Remarks . . . 84
3.7 Problems . . . 84

4 The Spectral Method **87**
4.1 Spectral Method Based on Fourier Series . . . 88
4.2 Spectral Method with Discrete Fourier Series . . . 93
4.3 Convergence Analysis . . . 97
4.4 Spectral Method Based on Chebyshev Polynomials . . . 102

4.5 Concluding Remarks 107
4.6 Problems 108

5 The Finite Element Method **113**
5.1 Strong Form, Weak or Variational Form, and Minimization 115
5.2 Discretization 117
5.3 More General Boundary Conditions 120
5.4 Sobolev Spaces 121
5.5 Convergence Analysis 125
5.6 Generalization to Two Dimensions 129
5.7 Where Are the Finite Elements? 136
5.8 Concluding Remarks 143
5.9 Problems 144

Bibliography **147**

Index **151**

Preface

This short book gives an introduction to numerical methods for elliptic partial differential equations (PDEs). It grew out of lecture notes Martin Gander prepared for a graduate course at McGill University in 2001 and 2002 which was followed by Felix Kwok as an undergraduate student, and both later taught the course at the University of Geneva (Martin Gander in 2004, 2008, and 2015 and Felix Kwok in 2012). The material is suitable for a one-semester course given to students from mathematics, computational science, and engineering, with two hours per week of lecturing and exercises.

This book is unique in three aspects:

1. It treats the four main discretization methods for elliptic PDEs, namely, finite difference methods, finite volume methods, spectral methods, and finite element methods.

2. It contains for each of these methods a complete convergence proof in the most simplified setting in order to illustrate the main analysis techniques needed to study these methods.

3. It contains runnable codes in MATLAB which give typical compact first implementations of these methods.

This allows the material to be taught with very little preparation, and all arguments are self-contained. It is also possible to study the material independently and individually, without taking a course. The book contains also an introduction to PDEs since often graduate students from various disciplines have not had such an introduction. Following the long tradition in numerical analysis in Geneva, the book is built on the historical development of the topics and contains precise descriptions of how methods and techniques were developed, including quotes from main contributors.

We are very thankful to Yves Courvoisier, who took notes in LaTeX for the first part of the course in 2008, when Martin Gander lectured in Geneva, and to Jérôme Michaud, who completed these notes when Felix Kwok lectured in Geneva in 2012. These lecture notes were the starting point of the book. Many thanks also to the many people who helped with the proofreading, in particular Laurence Halpern for the finite element chapter, Florence Hubert for the finite volume chapter, and Pratik Kumbhar, Parisa Mamouler, and Sandie Moody for the many misprints they found.

Martin J. Gander and Felix Kwok
August 2017

Chapter 1
Introduction

Les équations différentielles que nous avons démontrées, contiennent les résultats principaux de la théorie, elles expriment, de la manière la plus générale et la plus concise, les rapports nécessaires de l'analyse numérique avec une classe très-étendue de phénomènes, et réunissent pour toujours aux sciences mathématiques, une des branches les plus importantes de la philosophie naturelle.[a]

Joseph Fourier, Théorie analyique de la chaleur, 1822

All the mathematical sciences are founded on relations between physical laws and laws of numbers, so that the aim of exact science is to reduce the problems of nature to the determination of quantities by operations with numbers.

James C. Maxwell, Faraday's Lines of Force, 1855/1856

[a]The differential equations that we showed contain the main results of the theory; they represent in the most general and concise way the necessary relationship between numerical analysis and a wide class of natural phenomena, uniting one of the most important branches of natural philosophy forever to the mathematical sciences.

Ordinary and partial differential equations (ODEs and PDEs) are pervasive in scientific modeling today: Virtually every field in science and engineering uses differential equations, from biology (plant growth, population interactions, protein folding) to medicine (MRI, spread of infections, tumor growth), from physics (the big bang, black holes, particle accelerators) to chemistry (combustion, pollution, molecule interactions), and all engineering sciences (bridges, airplanes, cars, circuits, roads, etc.). There are myriad PDEs, and new ones are discovered by modeling every day. However, only a few of these PDEs have been analyzed mathematically, and even fewer possess closed-form solutions. With the advent of modern computers, solutions of PDEs can now be studied numerically. This book serves as an introduction to the four major techniques for doing so: *finite difference methods, finite volume methods, spectral methods, and finite element methods.* While for each of these techniques there exist excellent textbooks (see [25, 58, 36] for finite difference methods, [38, 17] for finite volume methods, [4, 61] for spectral methods, and [57, 32, 2] for the finite element method), a unified exposition introducing all these techniques is missing,[1] and this book attempts to fill this gap. It represents a one-semester course taught by the two authors at the universities of McGill and Geneva; throughout the book, Maple and MATLAB are used to provide hands-on experience with all the methods. At the end of each chapter, problems and projects are

[1]A very notable exception is the excellent book by Lui [41], which treats three of the four main methods.

included to help students familiarize themselves with the techniques learned and with the scientific computing tools Maple and MATLAB.

1.1 ▪ Notation

Throughout this book we use the following standard notation: For a real scalar function

$$y : \mathbb{R} \longrightarrow \mathbb{R}, \quad x \longmapsto y(x),$$

we denote its *derivatives* by

$$\frac{\mathrm{d}y(x)}{\mathrm{d}x} = y'(x), \quad \frac{\mathrm{d}^2y(x)}{\mathrm{d}x^2} = y''(x), \quad \dots.$$

The computer algebra system Maple handles derivatives using the function `diff`:

```
diff(y(x),x);
diff(y(x),x,x);
diff(y(x),x,x,x,x,x,x,x,x,x,x,x,x);
```

$$\frac{d}{dx}\,\mathrm{y}(x)$$

$$\frac{d^2}{dx^2}\,\mathrm{y}(x)$$

$$\frac{d^{12}}{dx^{12}}\,\mathrm{y}(x)$$

A more convenient way in Maple to denote higher derivatives is to use the *sequence operator $*,

```
diff(y(x),x$12);
```

$$\frac{d^{12}}{dx^{12}}\,\mathrm{y}(x)$$

which can be used in any other context as well since it simply constructs the associated sequence

```
x$12;
```

$$x, x, x, x, x, x, x, x, x, x, x, x$$

If the independent variable is t, then we often also use the notation

$$\frac{\mathrm{d}y(t)}{\mathrm{d}t} = \dot{y}(t), \quad \frac{\mathrm{d}^2y(t)}{\mathrm{d}t^2} = \ddot{y}(t), \quad \dots.$$

For a function depending on several variables, for example,

$$u : \mathbb{R}^3 \longrightarrow \mathbb{R}, \quad (x, y, z) \longmapsto u(x, y, z),$$

the *partial derivatives* with respect to each variable are denoted by[2]

$$\begin{aligned}
\frac{\partial}{\partial x}u(x,y,z) &= u_x(x,y,z) &= \partial_x u(x,y,z),\\
\frac{\partial^2}{\partial x^2}u(x,y,z) &= u_{xx}(x,y,z) &= \partial_{xx} u(x,y,z),\\
&\vdots &\vdots\\
\frac{\partial}{\partial y}u(x,y,z) &= u_y(x,y,z) &= \partial_y u(x,y,z),\\
&\vdots &\vdots\\
\frac{\partial}{\partial z}u(x,y,z) &= u_z(x,y,z) &= \partial_z u(x,y,z),\\
&\vdots &\vdots
\end{aligned}$$

Maple can also handle partial derivatives:

```
diff(u(x,y,z),x);
```

$$\frac{\partial}{\partial x}\,\mathrm{u}(x,\,y,\,z)$$

Additional interesting examples of derivatives using Maple are

```
diff(u(x,y,z),a);
```

$$0$$

```
ux:=diff(u(f(x),y,z),x);
```

$$ux := \mathrm{D}_1(u)(\mathrm{f}(x),\,y,\,z)\,\left(\frac{d}{dx}\,\mathrm{f}(x)\right)$$

```
f:=x->x^4;
ux;
```

$$f := x \to x^4$$

$$4\,\mathrm{D}_1(u)(x^4,\,y,\,z)\,x^3$$

```
u:=(x,y,z)->sin(x)*y+z^2;
ux;
```

$$u := (x,\,y,\,z) \to \sin(x)\,y + z^2$$

$$4\cos(x^4)\,y\,x^3$$

We see that Maple automatically substitutes variables with known expressions and computes the corresponding derivatives as soon as this is possible. We also see the operator `D` appear, which represents the following derivative operator:

```
fp:=D(f);
```

[2]This symbol for partial derivatives was introduced by Adrien-Marie Legendre in 1786.

$$fp := x \to 4\,x^3$$

```
D(sin);
D(sin+log);
```

$$\cos$$

$$\cos + \left(a \to \frac{1}{a}\right)$$

One can also use the derivative operator D for the partial derivatives

```
g:=(x,y)->x^2+y;
g1:=D[1](g);
g2:=D[2](g);
```

$$g := (x,\,y) \to x^2 + y$$

$$g1 := (x,\,y) \to 2\,x$$

$$g2 := 1$$

and all the results obtained are *Maple functions* that can be evaluated at a given argument:

```
g1(5,1);
```

$$10$$

There is a subtle but fundamental difference between the operator D and the function diff: The operator D acts on functions and returns functions, and the function diff acts on expressions and returns expressions. As an example, we obtain in Maple

```
f:=x->2*x^3;
diff(f(x),x);
unapply(diff(f(x),x),x);
D(f);
D(f)(x);
```

$$f := x \to 2\,x^3$$

$$6\,x^2$$

$$x \to 6\,x^2$$

$$x \to 6\,x^2$$

$$6\,x^2$$

which shows how to convert expressions into functions and vice versa. Finally, iterated application of the derivative operator D is obtained by

```
D(D(f));
(D@@2)(f);
```

$$x \to 12\,x$$

$$x \to 12\,x$$

One has to be careful when using the composition of functions, which is denoted by

```
D(sin@y);
```

$$(\cos@y)\,\mathrm{D}(y)$$

and not by

```
D(sin(y));
```

$$\mathrm{D}(\sin(y))$$

Definition 1.1 (gradient). *For a function $u : \mathbb{R}^d \to \mathbb{R}$, $d = 2, 3$, (e.g., $(x, y, z) \mapsto u(x, y, z)$), the* gradient $\nabla u : \mathbb{R}^d \to \mathbb{R}^d$ *is defined by*

$$\nabla u := \begin{cases} \begin{pmatrix} u_x \\ u_y \end{pmatrix} & \text{if } d = 2, \\ \begin{pmatrix} u_x \\ u_y \\ u_z \end{pmatrix} & \text{if } d = 3. \end{cases}$$

Definition 1.2 (divergence). *For a function $u : \mathbb{R}^d \to \mathbb{R}^d$, $d = 2, 3$, the* divergence $\nabla \cdot u : \mathbb{R}^3 \to \mathbb{R}$ *is defined by*

$$\nabla \cdot u := \begin{cases} \partial_x u_1 + \partial_y u_2 & \text{if } d = 2, \\ \partial_x u_1 + \partial_y u_2 + \partial_z u_3 & \text{if } d = 3. \end{cases}$$

Definition 1.3 (Laplacian). *For a function $u : \mathbb{R}^d \to \mathbb{R}$, $d = 2, 3$, the* Laplacian $\Delta u : \mathbb{R}^d \to \mathbb{R}$ *is defined by*

$$\Delta u := \nabla \cdot \nabla u = \begin{cases} \partial_{xx} u + \partial_{yy} u & \text{if } d = 2, \\ \partial_{xx} u + \partial_{yy} u + \partial_{zz} u & \text{if } d = 3. \end{cases}$$

Definition 1.4 (curl). *For a function $u : \mathbb{R}^3 \to \mathbb{R}^3$, the* curl $\nabla \times u : \mathbb{R}^3 \to \mathbb{R}^3$ *is defined by*

$$\nabla \times u = \begin{pmatrix} \partial_y u_3 - \partial_z u_2 \\ \partial_z u_1 - \partial_x u_3 \\ \partial_x u_2 - \partial_y u_1 \end{pmatrix}.$$

The curl is sometimes also used for a function $u : \mathbb{R}^2 \to \mathbb{R}^2$, in which case the result is just a scalar, namely, the third component in Definition 1.4. It can be interpreted as a vector orthogonal to the two-dimensional plane on which the function u is defined.

Maple knows these basic differential operators, and they are implemented (together with many other functions) in the `linalg` package:

```
with(linalg);
grad(x^2+3*y,[x,y]);
```

$$[2\,x \;\; 3]$$

```
diverge([x^2-3*y,4*x+y^5],[x,y]);
```

$$2\,x + 5\,y^4$$

```
diverge(grad(x^2+3*y,[x,y]),[x,y]);
laplacian(x^2+3*y,[x,y]);
```

$$2$$

$$2$$

```
curl([x^2-3*y+z,4*x+y^5-sin(z),x^2+y^2+z^2],[x,y,z]);
```

$$[2\,y + \cos(z) \;\; 1 - 2\,x, \; 7]$$

1.2 ▪ ODEs

An *ODE* is an equation whose solution depends on one variable only, and the solution appears together with its derivatives in the equation. Here are some examples:

$$y' = f(y), \quad \dot{u} = u^2 + t, \quad \ddot{x} = -x, \quad \ldots .$$

Each ODE containing higher-order derivatives can be reduced to a system of *first-order ODEs* by introducing additional variables, as we illustrate for the third example above, with the additional variable $y := \dot{x}$:

$$\ddot{x} = -x \quad \Longrightarrow \quad \begin{cases} \dot{x} = y, \\ \dot{y} = -x. \end{cases}$$

A differential equation admits in general a family of solutions; in order to select a unique solution from this family, one needs in addition an *initial condition*, which leads to the *initial value problem*

$$\begin{cases} y' &= f(y), \\ y(0) &= y^0. \end{cases}$$

ODEs are studied for their power of *modeling*. They allow us in many situations to easily obtain mathematical relations between quantities where such relations would be difficult (or impossible) to derive directly without derivatives. Many such laws were discovered by Newton and can be found in his monumental *Philosophiae Naturalis Principia Mathematica*; see Figure 1.1.

Example 1.5 (the pendulum). *Newton's second law of motion* says that the force F acting on an object is equal to the product of the mass m and the acceleration a (see Figure 1.2 from [45]):

$$F = m \cdot a.$$

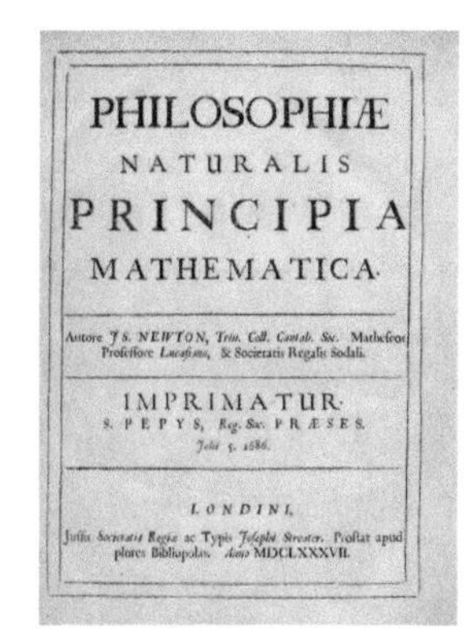

PHILOSOPHIÆ
NATURALIS
PRINCIPIA
MATHEMATICA.

Autore J S. NEWTON, Trin. Coll. Cantab. Soc. Matheseos Professore Lucasiano, & Societatis Regalis Sodali.

IMPRIMATUR.
S. PEPYS, Reg. Soc. PRÆSES.
Julii 5. 1686.

LONDINI,
Jussu Societatis Regiæ ac Typis Josephi Streater. Prostat apud plures Bibliopolas. Anno MDCLXXXVII.

Figure 1.1. *Isaac Newton* (25.12.1642–20.03.1726) *at the age of* 46 *and the title page of* Philosophiae Naturalis Principia Mathematica.

LEX II.

Mutationem motus proportionalem esse vi motrici impressæ, & fieri secundum lineam rectam qua vis illa imprimitur.

Figure 1.2. *Newton's second law of motion from the* Principia Mathematica *published in* 1686.

In the case of the pendulum, the only force acting on the pendulum is the *gravitational force*. The part relevant for the motion is the force orthogonal to the pendulum since the force in the direction of the pendulum is balanced by the tension of the string attaching it to the pivot. Thus, one has to split the total gravitational force into two components, as shown in Figure 1.3. The mass m is given, and the linear acceleration is obtained by multiplying the angular acceleration by the length of the string. Substituting leads to the differential equation

$$mg\sin(\theta) = -m\frac{\mathrm{d}^2(L\theta)}{\mathrm{d}t^2}.$$

The length of the string is constant, so we can take it out of the derivative. Simplifying the masses and isolating the second derivative on the left yields the more classical form

$$\ddot{\theta} = -\frac{g}{L}\sin(\theta). \tag{1.1}$$

This equation has no known closed-form solution. Maple attempts to find an implicit solution with the `dsolve` command:

```
dsolve(diff(theta(t),t,t)=-g/L*sin(theta(t)),theta(t));
```

$$\int^{\theta(t)} -\frac{L}{\sqrt{L\,(2\,g\cos(\_a) + \_C1\,L)}}\,d\_a - t - \_C2 = 0$$

$$\int^{\theta(t)} \frac{L}{\sqrt{L\,(2\,g\cos(\_a) + \_C1\,L)}}\,d\_a - t - \_C2 = 0$$

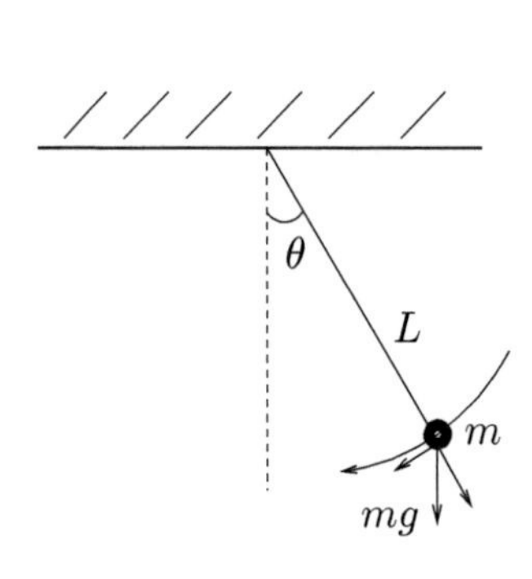

Figure 1.3. *The pendulum.*

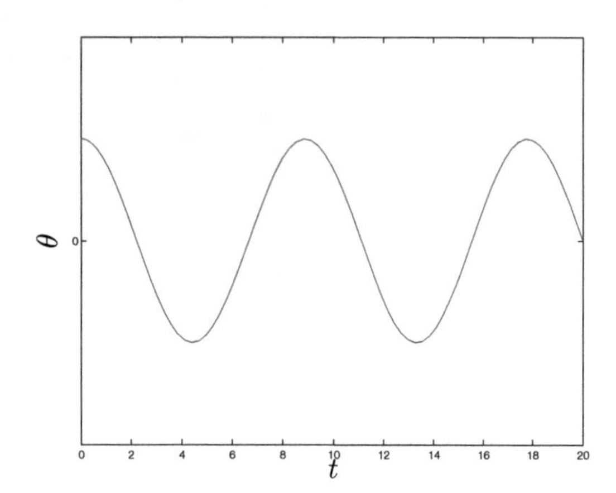

Figure 1.4. *Numerical solution of the nonlinear pendulum equation* (1.1).

A numerical method can be used to get an approximation of the solution. While Maple can also solve differential equations numerically (see Problem 1.3), MATLAB is our preferred tool for numerical calculations. In MATLAB, we obtain the numerical solution with the commands

```
f=@(t,y) [y(2); -1/2*sin(y(1))];
[t,y]=ode45(f,[0;20],[1/10; 0]);
plot(t,y(:,1));
xlabel('t');
ylabel('theta');
```

Note that we first transformed the second-order equation into a first-order system to define the right-hand side function `f` of the ODE. We then used the numerical ODE solver `ode45`, an adaptive *Runge–Kutta method* of order 4, together with the initial condition $\theta(0) = \frac{1}{10}$ and $\dot{\theta}(0) = 0$, to obtain the solution shown in Figure 1.4.

In order to obtain an analytical solution, we can approximate $\sin(\theta)$ for small θ by $\sin(\theta) \approx \theta$. Then the motion of this pendulum is for small θ approximately described by the ODE

$$\ddot{\theta} = -\frac{g}{L}\theta.$$

This equation has as solution a combination of the trigonometric functions

$$\sin(\sqrt{\tfrac{g}{L}}t), \quad \cos(\sqrt{\tfrac{g}{L}}t),$$

as one sees easily by inspection. The solution can also be obtained in Maple:

```
dsolve(diff(theta(t),t,t)=-g/L*theta(t),theta(t));
```

$$\theta(t) = \_C1 \sin\left(\frac{\sqrt{g}\,t}{\sqrt{L}}\right) + \_C2 \cos\left(\frac{\sqrt{g}\,t}{\sqrt{L}}\right)$$

Figure 1.5. *Left: Alfred J. Lotka* (2.3.1880–5.12.1949). *Courtesy of MetLife Archives. Right: Vito Volterra* (3.5.1860–11.10.1940).

Note how Maple denotes internal constants with an underscore. These constants are determined by suitable initial conditions, for example,

```
dsolve({diff(theta(t),t,t)=-g/L*theta(t),theta(0)=1/10,
  D(theta)(0)=0},theta(t));
```

$$\theta(t) = \frac{1}{10}\cos\left(\frac{\sqrt{g}\,t}{\sqrt{L}}\right)$$

Example 1.6 (population dynamics). A classical model of *predator-prey interaction* is the *Lotka–Volterra model*; see Figure 1.5 for a picture of the two inventors. Alfred J. Lotka, an American mathematician and physical chemist, arrived at the now classical model after being inspired by earlier studies of autocatalytic chemical reactions and of organic systems consisting of a plant species and a herbivorous animal species. We show in Figure 1.6 the specific situation Lotka considered in his book [40], where he first published his predator-prey model.

Independently, Vito Volterra did a statistical analysis of fish catches in the Adriatic Sea. As he tried to model the observations, he arrived at the same predator-prey system as Lotka in [63] in 1926; see Figure 1.7. There is still great interest in such ODE models (see, for example, the Swiss club of the "Volterriani," with an active journal at `http://web.liceomendrisio.ch/volterriani/volt.html`, which was founded by Antonio Steiner, who taught Martin Gander mathematics in high school in Solothurn).

We explain now the classical Lotka–Volterra model: Let $x(t)$ denote the *prey population* at time t and $y(t)$ be the corresponding *predator population*. Assuming that the prey population grows exponentially in the absence of predators, that the predator population diminishes exponentially without prey, and that the interaction is proportional to xy, representing a sort of probability that a predator finds a prey, we obtain the system of ODEs

$$\begin{cases} \dot{x} = ax - bxy, \\ \dot{y} = -cy + dxy, \end{cases} \tag{1.2}$$

where a, b, c, and d are positive constants. This system does not have a closed-form solution (for the closest one can get, see [55, 56]) and thus needs to be solved numerically; see Problem 1.7.

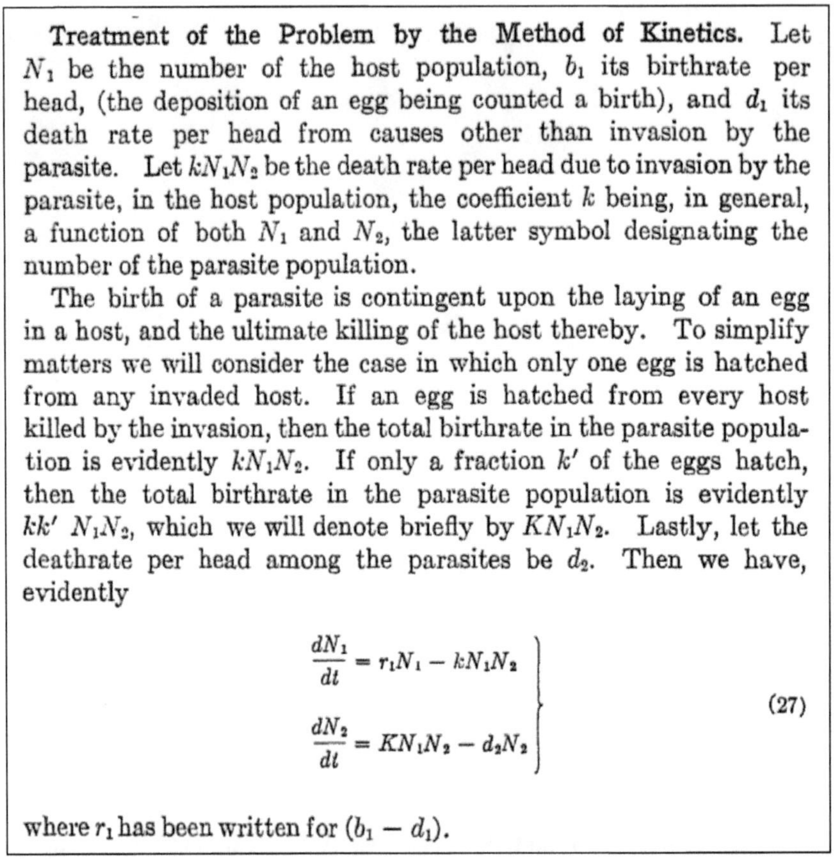

Treatment of the Problem by the Method of Kinetics. Let N_1 be the number of the host population, b_1 its birthrate per head, (the deposition of an egg being counted a birth), and d_1 its death rate per head from causes other than invasion by the parasite. Let kN_1N_2 be the death rate per head due to invasion by the parasite, in the host population, the coefficient k being, in general, a function of both N_1 and N_2, the latter symbol designating the number of the parasite population.

The birth of a parasite is contingent upon the laying of an egg in a host, and the ultimate killing of the host thereby. To simplify matters we will consider the case in which only one egg is hatched from any invaded host. If an egg is hatched from every host killed by the invasion, then the total birthrate in the parasite population is evidently kN_1N_2. If only a fraction k' of the eggs hatch, then the total birthrate in the parasite population is evidently $kk'\ N_1N_2$, which we will denote briefly by KN_1N_2. Lastly, let the deathrate per head among the parasites be d_2. Then we have, evidently

$$\left.\begin{aligned}\frac{dN_1}{dt} &= r_1N_1 - kN_1N_2\\ \frac{dN_2}{dt} &= KN_1N_2 - d_2N_2\end{aligned}\right\}\qquad(27)$$

where r_1 has been written for $(b_1 - d_1)$.

Figure 1.6. *Lotka invents the well-known Lotka–Volterra equations in* 1925. *Snapshot of page* 88 *of his book* [40]. *Reprinted with permission from Wolters Kluwer.*

Example 1.7 (crude oil production). Crude oil is an important commodity whose production has been tracked for decades; see the data in Figure 1.8, where we show past production worldwide and in a few countries (indicated with dots). The forecasting of future production rates based on economic considerations is of interest in many applications. A very simplistic model for the process of crude oil extraction can be obtained in the form of a system of ODEs. We denote by R the available reserves and by P the annual production. The resources diminish at the rate given by the production, so that $\dot{R} = -P$. The annual production rate P, however, fluctuates based on economic considerations. Let C_0 denote the cost of extracting oil, which can be modeled as

$$C_0 = b_1 - b_2R,$$

where $b_2 > 0$ indicates that it is cheaper to extract oil when the reserves R are plentiful. Now let x be the unit price of oil to the consumer. We assume that the demand for oil is slightly elastic; i.e., the demand D is given by

$$D(x) = D_0 - b_3x,$$

where $b_3 > 0$ is a small positive constant indicating that the demand drops slightly when oil prices are high. On the other hand, the supply tends to increase when oil companies

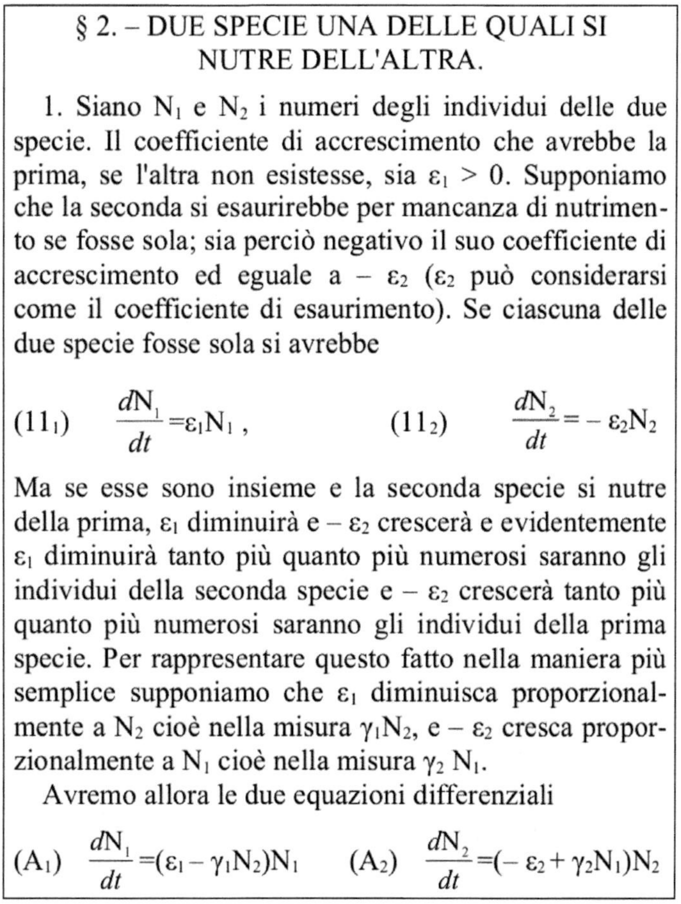

§ 2. – DUE SPECIE UNA DELLE QUALI SI NUTRE DELL'ALTRA.

1. Siano N_1 e N_2 i numeri degli individui delle due specie. Il coefficiente di accrescimento che avrebbe la prima, se l'altra non esistesse, sia $\varepsilon_1 > 0$. Supponiamo che la seconda si esaurirebbe per mancanza di nutrimento se fosse sola; sia perciò negativo il suo coefficiente di accrescimento ed eguale a $-\varepsilon_2$ (ε_2 può considerarsi come il coefficiente di esaurimento). Se ciascuna delle due specie fosse sola si avrebbe

$$(11_1) \quad \frac{dN_1}{dt} = \varepsilon_1 N_1, \qquad (11_2) \quad \frac{dN_2}{dt} = -\varepsilon_2 N_2$$

Ma se esse sono insieme e la seconda specie si nutre della prima, ε_1 diminuirà e $-\varepsilon_2$ crescerà e evidentemente ε_1 diminuirà tanto più quanto più numerosi saranno gli individui della seconda specie e $-\varepsilon_2$ crescerà tanto più quanto più numerosi saranno gli individui della prima specie. Per rappresentare questo fatto nella maniera più semplice supponiamo che ε_1 diminuisca proporzionalmente a N_2 cioè nella misura $\gamma_1 N_2$, e $-\varepsilon_2$ cresca proporzionalmente a N_1 cioè nella misura $\gamma_2 N_1$.

Avremo allora le due equazioni differenziali

$$(A_1) \quad \frac{dN_1}{dt} = (\varepsilon_1 - \gamma_1 N_2) N_1 \qquad (A_2) \quad \frac{dN_2}{dt} = (-\varepsilon_2 + \gamma_2 N_1) N_2$$

Figure 1.7. *Volterra invents the well-known Lotka–Volterra equations in* 1926. *Snapshot of page* 22 *of his book from* 1927. *Image courtesy of LiberLiber Association.*

stand to make a profit, so we assume that the supply S is given by $S(x) = S_0 + m(x - C_0)$. When the supply and demand are at equilibrium, i.e., when $S(x) = D(x)$, the equilibrium demand D^* is given by

$$D^* = \frac{b_3 S_0 + m(D_0 - b_1 b_3 + b_2 b_3 R)}{b_3 + m}.$$

On the one hand, when the equilibrium demand D^* exceeds production, the production should increase in order to satisfy demand; on the other hand, when $D^* - P < 0$, production should decrease. Thus, there is a positive constant c such that

$$\dot{P} = cP \cdot \operatorname{sgn}(D^* - P)|D^* - P|^{\alpha},$$

where α is an exponent to be determined. Redefining the undetermined constants if necessary, we arrive at our oil production model, which is given by the following system of ODEs:

$$\begin{cases} \dot{R} = -P, \\ \dot{P} = P \cdot \operatorname{sgn}(a_1 R - a_2 - a_3 P)|a_1 R - a_2 - a_3 P|^{a_4}, \end{cases}$$

where a_1, a_2, a_3, and a_4 are parameters to be matched with the available data.

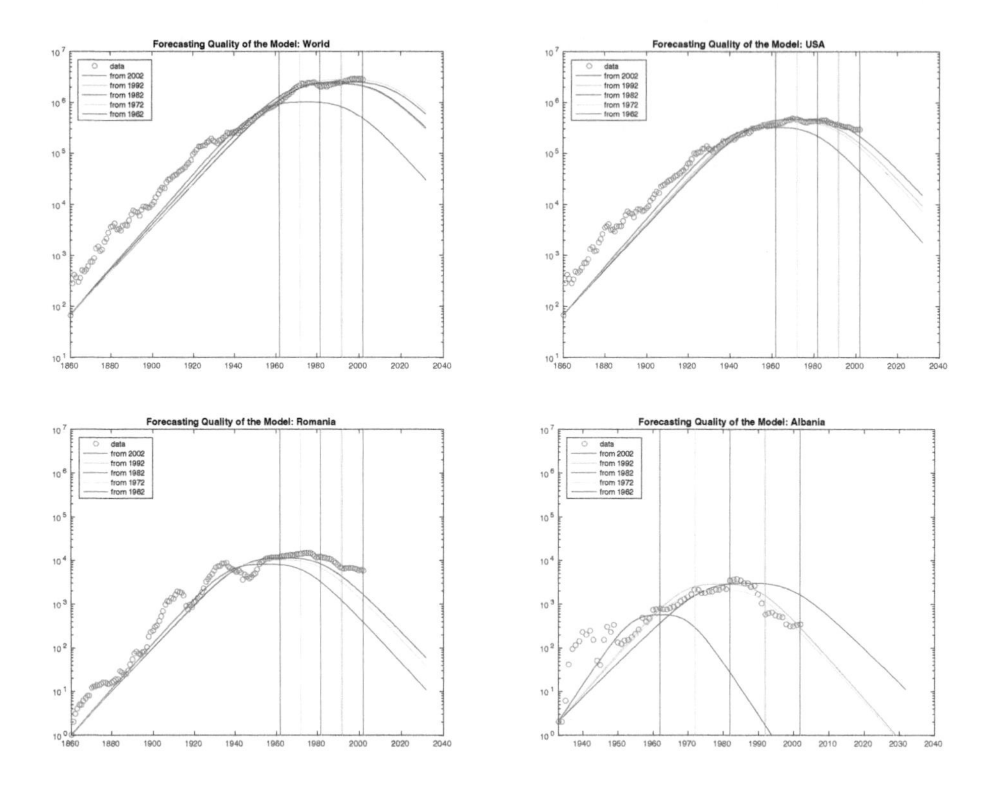

Figure 1.8. *Measured crude oil production worldwide and in the United States, Romania, and Albania (indicated by dots) and a simple ODE model with fitted parameters using various parts of the data to make predictions starting from a particular year.*

This model, simple as it is, still gives an adequate rough characterization of how oil production evolved over the last few decades, as one can see in Figure 1.8, where we show the production predicted by the model when the parameters are chosen to fit the available data. We now show how to obtain this fit using the following simple MATLAB commands. We first read and plot the data:

```
global yf df a x0;                                  % for easy use of MATLAB functions

d=[  67.00       283.00      412.00      363.00      299.00      353.00      508.00...
    478.00       521.00      599.00      746.00      743.00      893.00     1382.00...
   1513.00      1238.00     1299.00     1863.00     2137.00     2780.00     3603.00...
   3785.00      4150.00     3231.00     3352.00     3044.00     3894.00     3920.00...
   3878.00      4917.00     6390.00     7551.00     7079.00     6889.00     6067.00...
   7687.00      8946.00     9045.00     8537.00     8735.00     9808.00    10989.00...
  13500.00     15851.00    18636.00    21425.00    20695.00    27012.00    29795.00...
  31208.00     34393.00    37495.00    38132.00    43097.00    44784.00    48392.00...
  51971.00     59632.00    64223.00    72924.00    95198.00   103442.00   114814.00...
 135990.00    134742.00   140572.00   143067.00   162758.00   170150.00   190139.00...
 175817.00    165698.00   158274.00   174762.00   182904.00   199949.00   217893.00...
 250771.00    242190.00   253254.00   260954.00   268491.00   251924.00   276739.00...
 312250.00    330856.00   350362.00   384208.00   435290.00   427940.00   479064.00...
 543401.00    567525.00   596475.00   619488.00   689428.00   744197.00   774820.00...
 781706.00    836186.00   892119.00   940277.00  1013834.00  1081412.00  1167168.00...
1249662.00   1356040.00  1454391.00  1595145.00  1720448.00  1903813.00  2016375.00...
2126358.00   2330726.00  2312638.00  2212344.00  2430229.00  2408221.00  2412859.00...
```

```
2510271.00 2346078.00 2158408.00 2030681.00 2000912.00 2056640.00 2024621.00...
2161992.75 2174388.91 2266780.00 2334696.18 2403578.73 2431578.02 2500326.17...
2544941.95 2606937.54 2676219.52 2748182.47 2835994.67 2886370.65 2810290.33...
2895112.66 2845585.05 2756203.47]';

y=2002-length(d)+1:2002;                % corresponding year timeline
semilogy(y,d,'o')                       % plot data
xlabel('year');ylabel('oil production');
```

We introduce here the global variables `global yf df a x0;` which will simplify the use of optimization routines in MATLAB, as we will see. The data could also be read from a file using a command such as

```
d=load('World.txt');                    % load all data
```

where the file 'World.txt' should contain the numbers in ASCII text format line by line. Next, for a given set of the parameters a_j, we compute the solution of our differential equation using the MATLAB solver `ode45` and plot the solution together with the data:

```
a(1)=3.43456e-09;a(2)=0.8;a(3)=0;a(4)=2;
x0=[3*sum(d);d(1)];                     % choose initial value
[t,x]=ode45('F',y,x0);                  % solve model
semilogy(y,d,'o',t,x(:,2),'-');         % plot data
xlabel('year'); legend('Data','Fit','Location','SouthEast')
```

Here we used a function `F.m`, which was defined in a file with the same name, to implement the right-hand side of our model, namely,

```
function xp=F(t,x);
% F right hand side for simplistic crude oil forcasting model ODE
%   xp=F(t,x); right hand side function of the ODE x'=F(t,x,a), for a
%   simplistic crude oil forcasting model depending on the parameters
%   given in the global vector a

global a;

xp(1,1)=-x(2);
xp(2,1)=sign(a(1)*x(1)-a(2)-a(3)^2*x(2))*...
        abs(a(1)*x(1)-a(2)-a(3)^2*x(2))^a(4)*x(2);
```

For the initial conditions, we estimate the initial reserves (somewhat arbitrarily) to be three times the cumulative production over all the available data and the production to be equal to the first measured data point. While the parameter choice fits quite well, one might wonder if a better fit exists. We can try to minimize the distance between the data and the model using the function `Distance.m`, which is implemented as follows:

```
function d=Distance(aa)
% DISTANCE function to be used in fminsearch
%   d=Distance(aa) computes for a given parameter set in vector aa the
%   distance between the measured values df for the years yf contained
%   in global variables to the solution of the ODE model.

global yf df a x0;
```

```
a(1:length(aa))=aa;                     % put the current parameter choice
[t,x]=ode45('F',yf,x0);                 % solve the model with it
subplot(1,2,1)
semilogy(t,x(:,2),'-',yf,df,'o');       % plot data and current fit
subplot(1,2,2)
semilogy(yf,abs(x(:,2)-df),'-');        % plot the error
drawnow                                 % draw to see while optimizing
d=norm(x(:,2)-df,'inf');                % compute some norm of the error
```

The function `Distance.m` is meant to be used inside an optimization loop, with the model parameters as its only input arguments, while all other data are accessed via global variables. This way of passing data globally should not be used in real production code because it makes tracking of data access and modification difficult. Nonetheless, for very small prototype codes like ours, it is a slick way to proceed. We can now simply ask MATLAB to minimize the difference between the model and the data, measured in the infinity norm as specified in `Distance.m`, using the command

```
yf=y; df=d;
afit=fminsearch('Distance',[3.434e-09 0.7])
```

and watch how the Nelder–Mead algorithm implemented in `fminsearch` tries to minimize the distance. In our example, we only try to fit the first two parameters; the other ones remain at the value we set earlier. If the method has difficulties, as you can see in the plots produced, simply try a different initial guess; this is meant to be interactive!

To make a prediction, we can calculate the parameters using only part of the available data, i.e., cut off some of the years at the end and then verify how well the result fits the data we did not use:

```
ye=[0 10 20 30 40];                        % years to leave out at the end
for i=1:length(ye)
  yf=y(1:end-ye(i));                       % years to use for fitting
  df=d(1:length(yf));
  afit=fminsearch('Distance',[3.434e-09 0.7])
  a(1:length(afit))=afit;
  [tf{i},xf{i}]=ode45('F',[y y(end)+1:y(end)+30],x0);
end;
clf;
semilogy(y,d,'o',tf{1},xf{1}(:,2),'-r',tf{2},xf{2}(:,2),'-g',tf{3},...
  xf{3}(:,2),'-m',tf{4},xf{4}(:,2),'-c',tf{5},xf{5}(:,2),'-b');
legend('data','from 2002','from 1992','from 1982','from 1972',...
  'from 1962','Location','NorthWest')
line([2002 2002],[10 10^7],'Color','r');
line([1992 1992],[10 10^7],'Color','g');
line([1982 1982],[10 10^7],'Color','m');
line([1972 1972],[10 10^7],'Color','c');
line([1962 1962],[10 10^7],'Color','b');

title('Forecasting Quality of the Model: World');
```

This is how we computed the results shown in Figure 1.8. This way of fitting data is simple and very effective, and the commands above can be used for many other types of data fitting by simply changing the commands slightly.

As we have seen, most of the ODEs have no analytic solution, so one needs to resort to numerical methods to get an approximation of the solutions. A few classical methods are the forward and backward Euler method, Runge–Kutta methods, and linear multistep methods. For a comprehensive treatment of the numerical integration of ODEs, see the three authoritative monographs [26, 27, 28] that came out of the Geneva school of numerical analysis.

1.3 ▪ PDEs

Definition 1.8. *A* PDE *is a relation of the type*

$$F(x_1, \dots, x_n, u, u_{x_1}, \dots, u_{x_n}, u_{x_1 x_1}, u_{x_1 x_2}, \dots, u_{x_n x_n}, u_{x_1 x_1 x_1}, \dots) = 0, \quad (1.3)$$

where the unknown $u = u(x_1, \dots, x_n)$ *is a function of n variables and* $u_{x_j}, \dots, u_{x_i, x_j}, \dots$ *are its partial derivatives.*

Definition 1.9. *The highest order of differentiation occurring in a PDE is the* order of the PDE.

A more compact way of writing PDEs is to use the multi-index notation: Given a vector $\alpha = (\alpha_1, \dots, \alpha_n)$ of nonnegative indices, we write

$$D^\alpha(u) = \partial_{x_1}^{\alpha_1} \cdots \partial_{x_n}^{\alpha_n} u,$$

and we define the order of the multi-index to be $|\alpha| = \alpha_1 + \cdots + \alpha_n$. Then we can rewrite (1.8) as

$$F(\mathbf{x}, u, D^1 u, \dots, D^k u) = 0,$$

where k is the order of the PDE and $D^j u$ denotes the set of all partial derivatives of order j, $j = 1, \dots, k$.

Equation (1.3) is *linear* if F is linear with respect to u and all its derivatives; otherwise, it is *nonlinear*. There is a further, more refined classification of PDEs according to the criterion of linearity.

- *Linear PDEs* are equations of the form (1.3), where

$$F = \sum_{|\alpha| \le k} a_\alpha(\mathbf{x}) D^\alpha u - f(\mathbf{x});$$

  i.e., the coefficients multiplying $D^\alpha u$ are independent of u and its derivatives.

- *Semilinear PDEs* are equations where F has the form

$$F = \sum_{|\alpha| = k} a_\alpha(\mathbf{x}) D^\alpha u + f(\mathbf{x}, u, Du, \dots, D^{k-1} u);$$

  i.e., the coefficients multiplying the highest-order derivatives are independent of u and its derivatives.

- *Quasi-linear PDEs* are equations where F has the form

$$F = \sum_{|\alpha| = k} a_\alpha(\mathbf{x}, u, \dots, D^{k-1} u) D^\alpha u + f(\mathbf{x}, u, Du, \dots, D^{k-1} u);$$

  i.e., the coefficients multiplying the highest-order derivatives are independent of $D^k u$, where k is the order of the PDE.

- *Fully nonlinear PDEs* are equations where F is nonlinear with respect of the highest order derivatives of u.

For semilinear second-order equations of the form

$$au_{xx} + bu_{xy} + cu_{yy} = f(x, y, u, u_x, u_y), \tag{1.4}$$

there is also a classification based on the discriminant $b^2 - 4ac$:

- $b^2 < 4ac$: elliptic PDE, e.g., a Poisson equation $u_{xx} + u_{yy} = f(x, y)$;
- $b^2 > 4ac$: hyperbolic PDE, e.g., wave equation $u_{tt} - \frac{1}{c^2}u_{xx} = 0$;
- $b^2 = 4ac$: parabolic PDE, e.g., (advection-) diffusion $u_t + cu_x = \kappa u_{xx}$.

If there are more than two independent variables, (1.4) becomes

$$\sum_{j=1}^{m}\sum_{k=1}^{m} a_{jk}u_{x_j x_k} = f(x_1, \ldots, x_m, u, u_{x_1}, \ldots, u_{x_m}), \tag{1.5}$$

where we assume without loss of generality that $a_{jk} = a_{kj}$ when $k \neq j$ because for smooth enough functions, mixed partial derivatives are equal regardless of the order in which they are taken. In this case, the classification into elliptic, parabolic, and hyperbolic PDEs can be generalized by looking at the eigenvalues of the matrix $A := [a_{jk}]$:

- The PDE is *elliptic* if all eigenvalues of A are of the same sign and none vanishes.
- The PDE is *parabolic* if all eigenvalues of A are of the same sign except one, which vanishes.
- The PDE is *hyperbolic* if the matrix A is nonsingular and if all eigenvalues of A are of the same sign except one, which has *the opposite sign*.

Here are some examples of PDEs:

$$\Delta u = f,\ u_t + \Delta u = f,\ u_{xx} - u_{yy} = u_{tt},\ u_t + a \cdot \nabla u - \nu\Delta u = f, \ldots.$$

Some of these equations describe interesting physical phenomena (first and last one), whereas others are not interesting or might not even have a solution. To obtain a unique solution, we must specify a physical domain on which we consider the equation, $\Omega \subset \mathbb{R}^d$, $d = 1, \ldots, 3$, and add appropriate boundary conditions and also initial conditions if the equation depends on time. As an example, for the *Poisson equation*, we could consider

$$\left\{\begin{array}{rcll} -\Delta u &=& f & \Omega := (0,1) \times (0,1), \\ u &=& 0 & \partial\Omega, \end{array}\right.$$

where $\partial\Omega$ denotes the boundary of the domain Ω.

Another example is the *heat equation*. The heat in a one-dimensional object, $\Omega := (0, 1)$, is described by the PDE

$$u_t = u_{xx}.$$

To get a unique solution, one needs to specify boundary conditions at both ends and also an initial condition, which is the initial temperature of the object, for example,

$$\left\{\begin{array}{rcll} u_t &=& u_{xx} & \text{in } \Omega \times (0, T), \\ u(0, t) &=& 1, & \\ u(1, t) &=& 0, & \\ u(x, 0) &=& e^{-x}. & \end{array}\right. \tag{1.6}$$

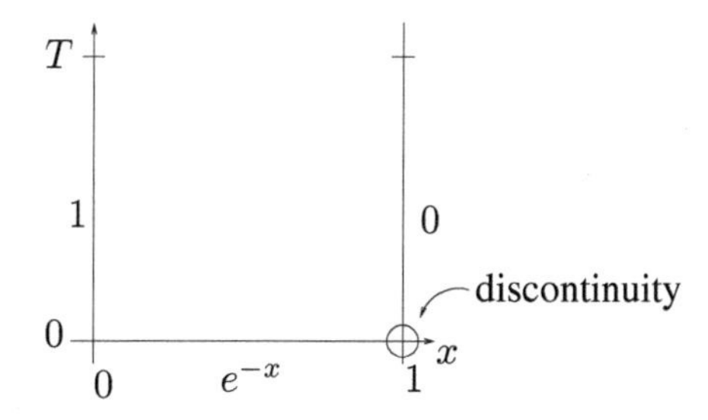

Figure 1.9. *Figure of the domain* $\Omega \times (0, T)$ *over which the heat equation is defined in* (1.6), *together with the initial and boundary conditions.*

A drawing often helps to understand the problem setting and can quickly point out potential difficulties, for example, the discontinuity at the right end between the initial and boundary condition, as shown in Figure 1.9.

There are very few PDEs with closed-form solutions. Also, the theoretical study of the existence and uniqueness of solutions for certain equations is currently an active research area. For a few fundamental PDEs, complete results are available, but already for the Navier–Stokes equations, which we will encounter later, the existence and uniqueness question in three dimensions has not been fully answered and is currently one of the seven mathematical problems that carry a prize of US $\$1$ million.[3] PDEs, rather than ODEs, permit a much larger range of problems to be modeled, and a large part of numerical analysis research today is devoted to the numerical study of PDEs.

1.4 ▪ The Heat Equation

Newton's law of cooling states that the variation of temperature of a body is proportional to the difference between its own temperature and the temperature of the surrounding environment.[4] If u denotes the temperature of the body, then Newton's law states that

$$u_t = k(\tilde{u} - u),$$

where $\tilde{u}$ is the ambient temperature and k is a constant of proportionality known as the heat transfer coefficient. This ODE is a rather simple model for the evolution of the temperature of the body under the influence of an external temperature. It assumes that the body is very simple and behaves like a single point in space. If the body is made of different materials so that some parts react differently to the external temperature, we need to add the dependence of the temperature on the spatial location in order to obtain a more accurate model.

Fourier's law of heat flux states that the flow of heat is from hot to cold; i.e., it is proportional to the negative gradient of the temperature (see Figure 1.10 for the original statement by Fourier),

$$\mathbf{F} = -a\nabla u,$$

where $\mathbf{F}$ is the vector field describing the flow of the temperature and a is a constant of proportionality that depends on the thermal conductivity and the specific heat capacity of the material. However, this constant can now depend on the spatial location if the material properties are not uniform. Some proportionality constants are listed in Table 1.1. Note also the minus sign in the formula for the heat flux: The flux is in the opposite

[3]See the seven Millennium Prize Problems at `www.claymath.org/millennium-problems`.

[4]It is not completely clear how this law was established historically; see the fascinating description in [51].

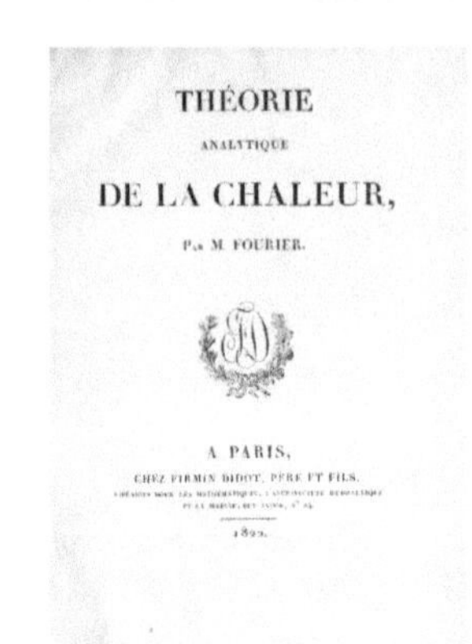

THÉORIE

ANALYTIQUE

DE LA CHALEUR,

PAR M. FOURIER.

A PARIS,

CHEZ FIRMIN DIDOT, PÈRE ET FILS,

1822.

131. Expression analytique du flux dans l'intérieur d'un solide quelconque. L'équation des températures étant $v = f(x, y, z, t)$ la fonction $-K\omega . \frac{dv}{dz}$ exprime la quantité de chaleur qui traverse, pendant l'instant dt, une aire infiniment petite ω, perpendiculaire à l'axe des z, au point dont les coordonnées sont x, y, z, et dont la température est v après le temps écoulé t.

Figure 1.10. *Joseph Fourier* (21.03.1768–16.05.1830) *discovers the law of heat flux; shown is copy of the very nice table of contents entry in* [19, p. 611], *which gives a summary, not just a title.*

Table 1.1. *Table of the proportionality constants of some different materials.*

Environment	a [m^2/sec]
water	0.00144
granite	0.011
iron	0.12
aluminum	0.86
silver	1.71

direction of the gradient of the temperature since the gradient points in the direction of maximum temperature increase.

To derive a PDE describing the temperature evolution over time in a domain $\Omega \subset \mathbb{R}^3$, we look at the total amount of heat in the domain at time t, which is

$$\int_\Omega u(\mathbf{x}, t) d\mathbf{x},$$

and describe how this quantity varies over time. The change in the total amount of heat in Ω is given by the flow of heat across the boundary of Ω and possible heat sources or sinks in the domain:

$$\frac{\partial}{\partial t} \int_\Omega u(\mathbf{x}, t) d\mathbf{x} = \int_{\partial\Omega} -\mathbf{F}(\mathbf{x}(s), t) \cdot \mathbf{n}(s) ds + \int_\Omega f(\mathbf{x}, t) d\mathbf{x}. \tag{1.7}$$

Here the function f denotes the heat sources or sinks, and $\mathbf{n}(s)$ is the unit outward normal to the boundary $\partial\Omega$ of Ω. Thus, $\mathbf{F}(\mathbf{x}(s), t) \cdot \mathbf{n}(s)$ is precisely the heat flux leaving

134. Il est facile de déduire du théorème précédent, l'équation générale du mouvement de la chaleur, qui est

$$\frac{dv}{dt} = \frac{K}{C.D} \cdot \left(\frac{d^2v}{dx^2} + \frac{d^2v}{dy^2} + \frac{d^2v}{dz^2}\right). \qquad \text{(AE)}$$

Figure 1.11. *Copy of the table of contents entry in* [19, p. 612].

the domain across the boundary, which explains the minus sign. The tangential part of the heat flux at the boundary does not contribute to the change of the total heat since it neither leaves nor enters the domain. Using Fourier's heat law and the divergence theorem, we obtain

$$\begin{aligned}\frac{\partial}{\partial t}\int_\Omega u(\mathbf{x},t)d\mathbf{x} &= \int_{\partial\Omega} -\mathbf{F}\cdot\mathbf{n}ds + \int_\Omega f(\mathbf{x},t)d\mathbf{x}\\ &= \int_{\partial\Omega} a\nabla u\cdot\mathbf{n}ds + \int_\Omega f(\mathbf{x},t)d\mathbf{x}\\ &= \int_\Omega \nabla\cdot a\nabla u d\mathbf{x} + \int_\Omega f(\mathbf{x},t)d\mathbf{x}.\end{aligned}$$

Since in general, Ω does not depend on time, we can change the order of integration and differentiation, and obtain

$$\int_\Omega \left(\frac{\partial}{\partial t}u - \nabla\cdot a\nabla u - f\right)(x,t)d\mathbf{x} = 0.$$

If the function under the integral is continuous, then varying the arbitrary domain of integration Ω allows us to conclude that the temperature u satisfies the PDE

$$\frac{\partial}{\partial t}u - \nabla\cdot a\nabla u = f.$$

This equation, which is parabolic according to our classification, is called the *heat equation* or *diffusion equation*. It appears as the next entry after Fourier's law of heat flux in [19]; see Figure 1.11 for the original statement by Fourier in 1822, who derived the equation by arguing at the differential level ("on voit par-là qu'il s'accumule durant l'instant dt dans l'intérieur de cette molécule, une quantité totale de chaleur égale à $K(\frac{d^2v}{dx^2}+\frac{d^2v}{dy^2}+\frac{d^2v}{dz^2})$"). Fourier had essentially obtained these results already in a groundbreaking manuscript submitted to the French Academy in 1807, but the appointed committee of reviewers (Laplace, Lagrange, Monge, and Lacroix) did not appreciate the proposed way of solving the equation using infinite series; see [44].

Example 1.10. Suppose a nail of length L with initial temperature $u_0(x)$, $x \in (0,L)$, is placed between two ice cubes, as shown in Figure 1.12.[5] We assume that the nail is thin enough so that we can consider it to be one dimensional. The ice cubes are naturally at temperature zero, and they touch the nail at both extremities. This defines the boundary conditions of the problem, and for the equation describing the temperature distribution

[5]This example is a variation of the original example of Fourier, which led to the discovery of Fourier series [19, p. 161]: "Ainsi la question actuelle consiste à déterminer les températures permanentes d'un solide rectangulaire infini, compris entre deux masses de glace B et C et une masse d'eau bouillante A."

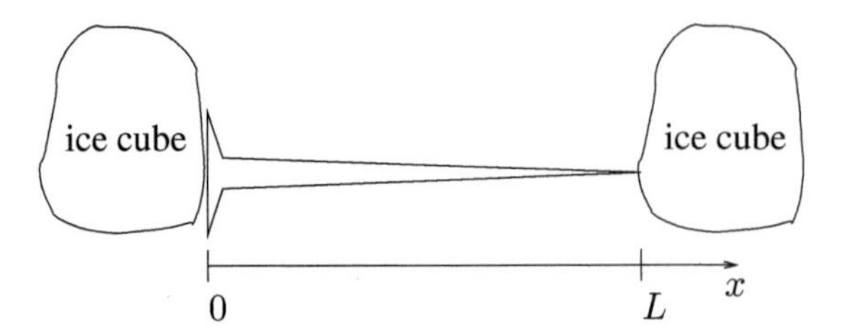

Figure 1.12. *Figure of a nail cooled by two ice cubes placed on either side of it.*

in the nail we obtain

$$\begin{cases} u_t &= \nu u_{xx} & \text{in } (0,L)\times(0,1), \\ u(x,0) &= u_0(x) & \text{in } \Omega, \\ u(0,t) &= 0 & \text{in } (0,T), \\ u(L,t) &= 0 & \text{in } (0,T), \end{cases} \tag{1.8}$$

where ν is a positive real constant. The temperature of the nail is only influenced by the ice cubes at both ends; there is no other heat source or sink in this example, which explains why there is no source term f in the equation.

This is one of the few PDEs that have an analytic solution. Following in the footsteps of Fourier, we construct a solution by *separation of variables*. The main idea behind this method is first to seek simple solutions of the PDE and then to use the linearity of the equation to construct more general solutions. We suppose that the temperature u is a product of two functions,[6] one depending on x and the other on t:

$$u(x,t) = a(x)b(t).$$

Inserting this expression into (1.8) gives the relation

$$a(x)b'(t) = \nu a''(x)b(t).$$

Rearranging so that the left-hand side contains only functions of x and the right-hand side only functions of t leads to

$$\nu\frac{a''(x)}{a(x)} = \frac{b'(t)}{b(t)}.$$

Since equality must hold for all x and t, the expressions on each side of the equal sign must be constant. Denoting this constant by σ, we obtain the two equations

$$\begin{cases} a'' &= \frac{\sigma}{\nu}a, \\ b' &= \sigma b. \end{cases} \tag{1.9}$$

This system we obtained from the heat equation consists now of ODEs, and those need to be completed with initial (or boundary) conditions. Using the boundary conditions from the heat equation (1.8), we obtain at $x = 0$

$$u(0,t) = a(0)b(t) = 0.$$

Thus, one or both terms must be zero. If b is identically zero, the solution u is zero. This is not desirable since it does not verify the initial condition. We thus set $a(0)$ to be

[6]Fourier [19, p. 163]: "Nous examinerons d'abord si la valeur de v peut être représentée par un pareil produit."

zero and hence have obtained an initial condition for $a(x)$. We need, however, a second condition to solve the corresponding second-order ODE. Taking the second boundary condition of the heat equation, we get

$$u(L,t) = a(L)b(t) = 0.$$

Again considering b to be zero leads nowhere, and thus we set $a(L) = 0$. Now the second-order equation for $a(x)$ is complete, with one initial and one final condition (or two boundary conditions),

$$\begin{cases} a'' &= \frac{\sigma}{\nu}a \quad \text{in } (0,L), \\ a(0) &= 0, \\ a(L) &= 0. \end{cases}$$

Depending on σ, we need to consider three different cases:

1. ($\sigma = 0$). If σ is zero, the solution is a linear function,

   $$a(x) = cx + d.$$

   Using the boundary conditions, we find that the constants c and d are zero, and we obtain again the zero solution, which is of no interest.

2. ($\sigma > 0$). A positive σ leads to an exponential solution of the form

   $$a(x) = ce^{\sqrt{\frac{\sigma}{\nu}}x} + de^{-\sqrt{\frac{\sigma}{\nu}}x}.$$

   The boundary conditions lead to the system of linear equations

   $$\begin{cases} c + d = 0, \\ ce^{\sqrt{\frac{\sigma}{\nu}}L} + de^{-\sqrt{\frac{\sigma}{\nu}}L} = 0, \end{cases}$$

   which has again only the solution $c = 0$ and $d = 0$. Thus, we only obtain identically zero solutions for the functions a and u.

3. ($\sigma < 0$). In that case, the unknown a is of the form

   $$a(x) = c\cos\left(\sqrt{\frac{-\sigma}{\nu}}x\right) + d\sin\left(\sqrt{\frac{-\sigma}{\nu}}x\right).$$

   We denote by λ the constant appearing in the cosine and sine, i.e., $\lambda = \sqrt{\frac{-\sigma}{\nu}}$. Using the boundary condition at $x = 0$, we find again

   $$a(0) = c = 0,$$

   but at $x = L$ the boundary condition gives

   $$a(L) = d\sin(\lambda L) = 0,$$

   and we can have a nonzero constant d since $\sin(\lambda L)$ is zero for

   $$\lambda = \frac{n\pi}{L}, \quad n = 0, 1, 2, \dots.$$

   For each integer n, we thus have the solution

   $$a_n(x) = d_n \sin\left(\frac{n\pi}{L}x\right),$$

   and the corresponding constant σ must take on the value $\sigma_n = -\nu(\frac{n\pi}{L})^2$.

For each of those solutions a_n, we look for the corresponding function $b_n(t)$. Equation (1.9) defining b becomes an equation for b_n,

$$b_n' = -\nu \left(\frac{n\pi}{L}\right)^2 b_n,$$

with solution

$$b_n = c_n e^{-\nu(\frac{n\pi}{L})^2 t}.$$

Multiplying both solutions a_n and b_n gives

$$u_n(x,t) = c_n e^{-\nu(\frac{n\pi}{L})^2 t} \sin\left(\frac{n\pi}{L}x\right), \quad n = 0, 1, 2, \ldots,$$

which are all solutions of the heat equation with homogeneous boundary conditions

$$\begin{cases} \dfrac{\partial}{\partial t} u_n = \nu \dfrac{\partial^2}{\partial x^2} u_n, \\ u_n(0,t) = u_n(L,t) = 0. \end{cases}$$

The sum of all such solutions u_n is, by linearity, also solution to this problem,

$$u(x,t) = \sum_{n=1}^{\infty} u_n(x,t) = \sum_{n=1}^{\infty} c_n e^{-\nu(\frac{n\pi}{L})^2 t} \sin\left(\frac{n\pi}{L}x\right).$$

The only condition of our original problem we have not yet satisfied is the initial condition.[7] Our series solution $u(x,t)$ at $t = 0$ is

$$u(x,0) = \sum_{n=1}^{\infty} c_n \sin\left(\frac{n\pi}{L}x\right) = u_0(x).$$

Thus, we must find constants c_n so that $u(x,0) = u_0(x)$, the initial condition. But $u(x,0)$ is simply a *Fourier series*,[8] and thus the coefficients c_n must be the Fourier coefficients of the initial condition u_0,

$$c_n = \frac{2}{L} \int_0^L \sin\left(\frac{n\pi}{L}x\right) u_0(x) dx.$$

Remark 1.1. *It is not a coincidence that we obtained a Fourier series since second-order operators generally possess eigenfunctions which form an orthonormal system, a result due to Sturm and Liouville.*

Remark 1.2. *Our closed-form solution reveals an interesting property of solutions of the heat equation in general: The Fourier coefficients c_n are multiplied by the exponential term $e^{-\nu(\frac{n\pi}{L})^2 t}$, which depends strongly on n. The bigger n is, the faster the exponential term goes to zero. This means that if one starts with an initial condition that contains high-frequency components, n large, those will disappear quickly, and*

[7] Fourier [19, p. 164]: "Il reste à remplir une troisième condition."

[8] The search for a solution of the heat equation (Fourier considered the steady case, a two-dimensional Laplace equation) led Fourier to his major invention of the Fourier series [19, p. 165]: "On pourrait douter qu'il existât une pareille fonction, mais cette question sera pleinement éclaircie par la suite." The interested reader is invited to continue reading the courageous approach of Fourier (pp. 165ff), with the key idea finally coming on page 232.

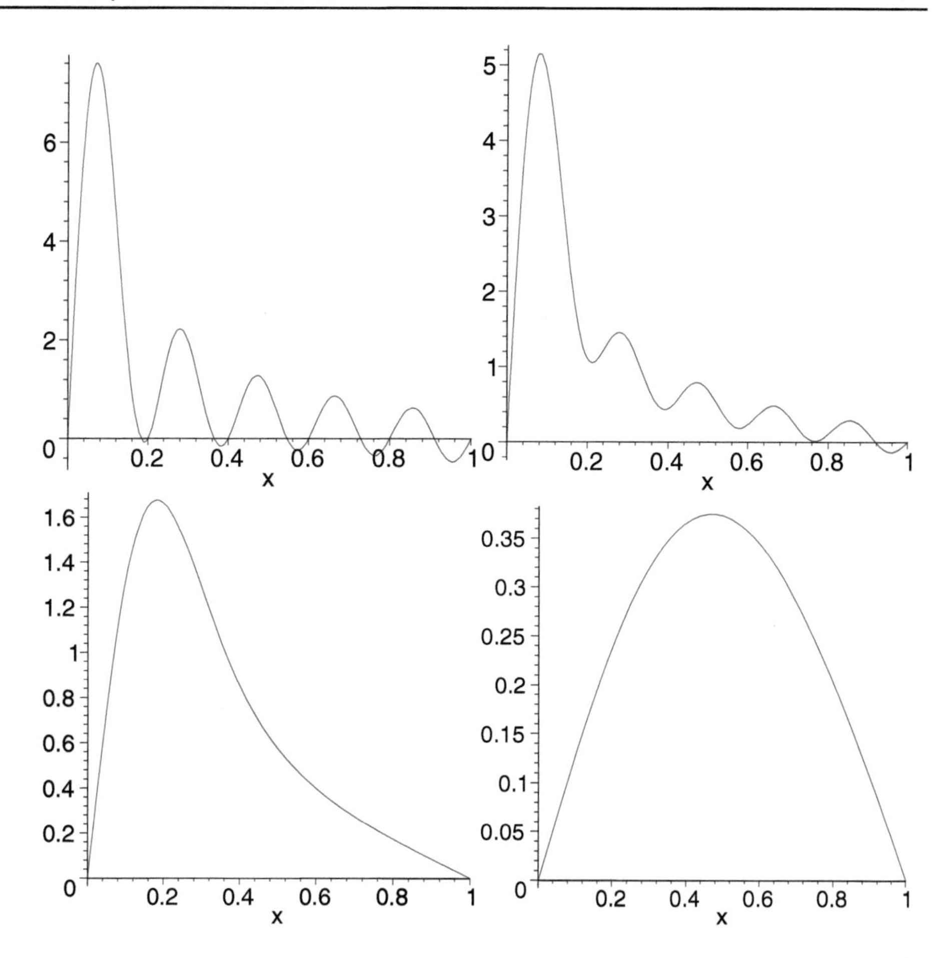

Figure 1.13. *Solution of the one-dimensional heat equation: Initially, at time $t = 0$, one can see many high-frequency components in the initial condition. Those quickly disappear from the solution, as one can see at times $t = 0.001$, $t = 0.01$, and $t = 0.1$.*

only the low frequency components, n small, will remain after a short period of time. This can be easily seen in Maple:

```
u:=sum(c[n]*sin(n*Pi/L*x)*exp(-nu*(n*Pi/L)^2*t),n=1..N);
nu:=1;L:=1;
N:=10;
c:=[1$N];
u;
t:=0;plot(u,x=0..L);
t:=0.001;plot(u,x=0..L);
t:=0.01;plot(u,x=0..L);
t:=0.1;plot(u,x=0..L);

t:='t';plot3d(u,x=0..1,t=0..0.1,axes=boxed);
```

We show in Figure 1.13 *the four results obtained at different times and in Figure* 1.14 *the evolution in space-time of the solution.*

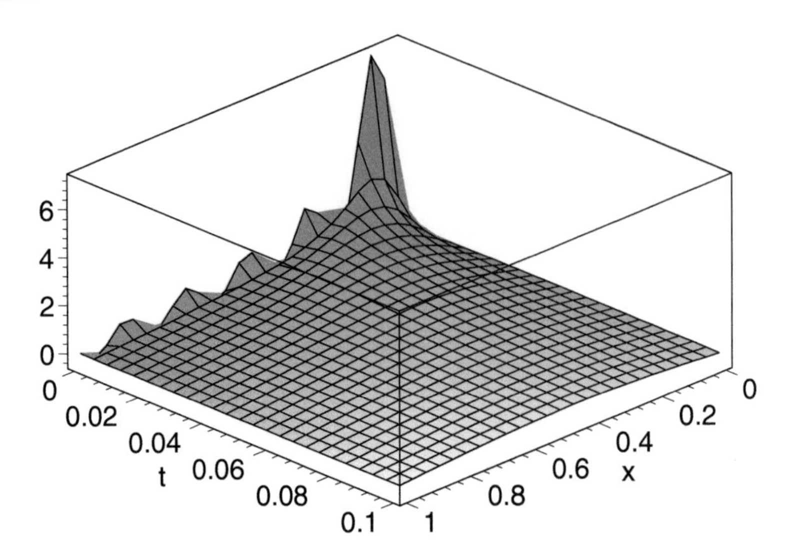

Figure 1.14. *Solution of the one-dimensional heat equation in space-time, where one can clearly see how high-frequency components are quickly attenuated in the solution.*

Remark 1.3. *Each term u_n of the solution decreases when t increases because of the exponential term $e^{-\nu(\frac{n\pi}{L})^2 t}$. Thus, the general solution $u(x,t)$ tends to zero when t tends to infinity,*

$$u(x,t) \to 0, \quad (t \to \infty).$$

This means in the context of our problem in Figure 1.12 that the nail cools until it reaches the temperature of the ice cubes.

Example 1.11. A more concrete example is the temperature distribution in the living room of Martin Gander, when he was living in Montreal, where in winter the outdoor temperature regularly drops below $-20°$C. The living room in apartment 206 on 3421 Durocher Street is shown in Figure 1.15. One can see the entrance door in the top left corner, the angle couch in the top right corner, the desk on the right below, and the dining table directly in front of the window at the bottom. The red line indicates the location of the heater in the room. The window insulation was very modest, as the two glass windows regularly froze from the interior. The entrance door was connected to a hallway, which was regularly heated to about $15°$C. We assume that the walls are perfectly insulated and thus impose homogeneous Neumann conditions at the boundaries represented by walls, so that the normal derivative, which represents the heat flux, is zero. At the door and the window, we assume that the temperature equals the outside environment and thus impose Dirichlet conditions with the corresponding temperatures.

1. If we are interested in the evolution of the temperature, then the temperature distribution in the living room is obtained from the *heat equation*

$$u_t = \nu\Delta u + f.$$

Clearly, a solution using a series expansion is difficult here because of the complex geometry, so one needs to use numerical methods. An example of the temperature evolution using a finite difference method is given in Figure 1.16, which represents the situation when one returns from a skiing weekend during which

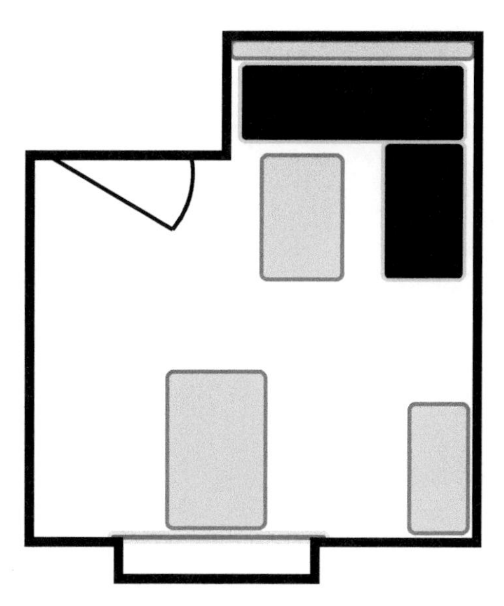

Figure 1.15. *Living room of Martin Gander's apartment on Durocher Street in Montreal.*

the heater was turned off. One can see how the room heats up after the heater has been turned on and in particular how the temperature is increasing the most slowly in the area around the couch. This phenomenon, which we felt regularly, is the reason why we usually avoided the couch area after returning.

2. If the heating is left on, the solution of the heat equation eventually becomes stationary, as one can see as one moves from the top left to the bottom right of Figure 1.16. Mathematically, this means that the source is independent of time, i.e., $f = f(x)$, and we obtain a stationary temperature distribution when $t \to \infty$:

$$\frac{\partial u}{\partial t} \to 0, \quad t \to \infty.$$

The limiting heat distribution then satisfies the *elliptic* PDE

$$-\nu \Delta u = f$$

with the same boundary conditions. This is the *Poisson equation*.

3. If there is no heater, then the source is the zero function $f = 0$ and the equation becomes the *Laplace equation*,

$$\Delta u = 0,$$

again with the same boundary conditions, whose solution is shown in the top left frame in Figure 1.16, before the heater was turned on.

1.5 ▪ The Advection-Reaction-Diffusion Equation

We denote by u the concentration of a certain substance in a fluid in motion, whose flow follows a given velocity vector field $\mathbf{a}$, as shown in Figure 1.17. We assume that

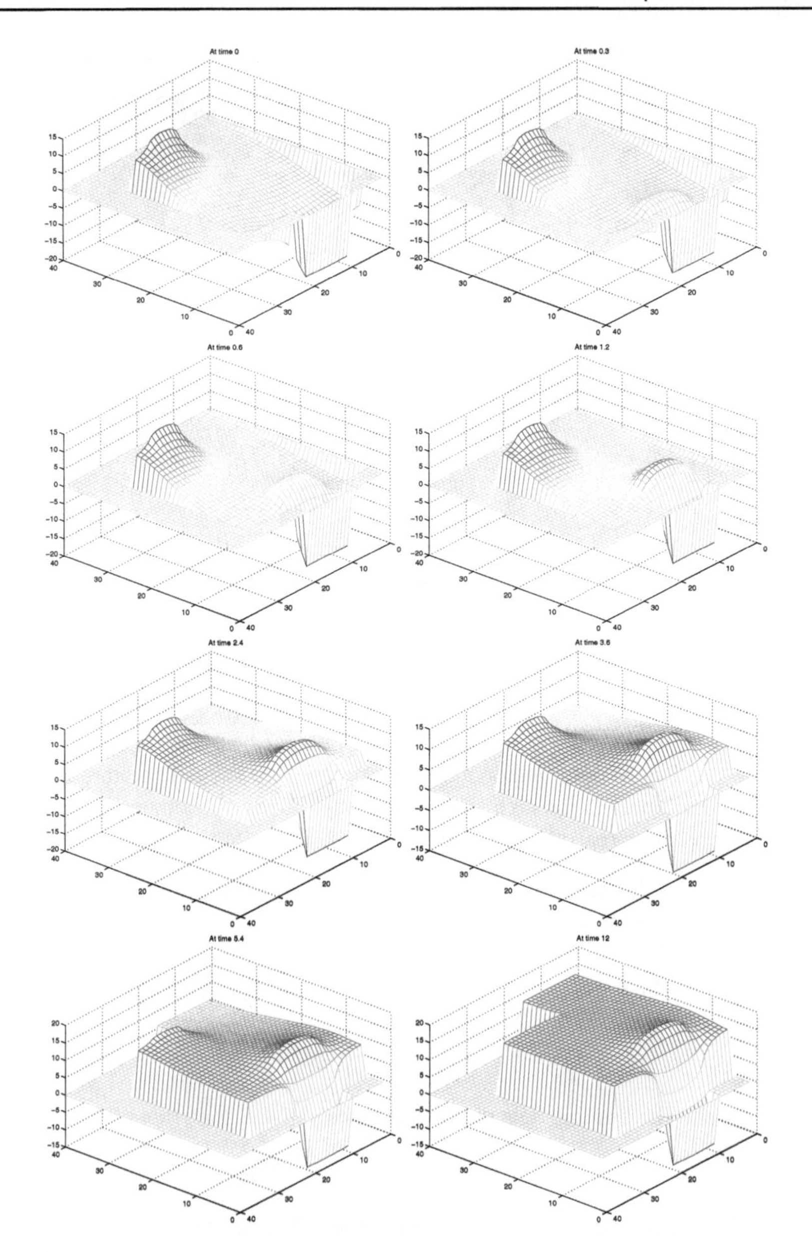

Figure 1.16. *Evolution of the temperature after returning from a skiing weekend.*

the substance is added in a certain location and also that it can chemically react with its environment, both of which are modeled by the possibly nonlinear function $f(u, x, t)$. Here we adopt the convention that $f > 0$ means that substance is created and $f < 0$ means that substance disappears.

The flow of the substance in the fluid has two components: The first is naturally the influence of the motion of the fluid itself, which simply transports the substance according to the vector field **a**, and the second is the diffusion of the substance in the fluid, which leads to the total flux of the substance,

$$\mathbf{F} = \mathbf{a}u - \nu\nabla u.$$

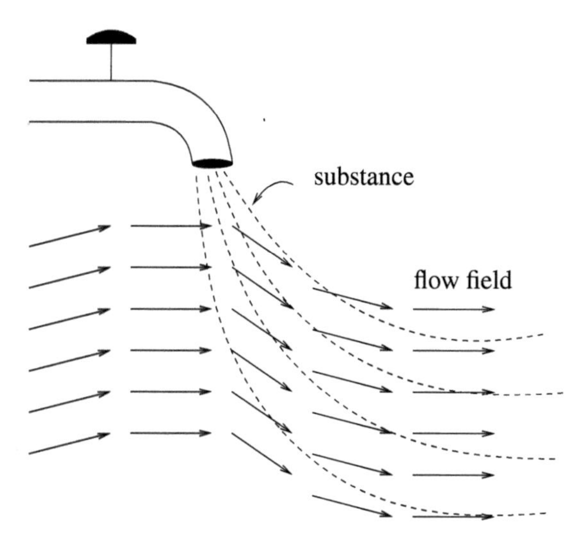

Figure 1.17. *Drawing showing the mixing of a substance in a moving fluid with a concentration varying because of chemical reactions creating or eliminating the considered substance.*

As for the heat equation, using conservation of the substance in an arbitrary domain Ω, we obtain the PDE

$$\begin{aligned} u_t &= -\nabla \cdot \mathbf{F} + f \\ &= \nabla \cdot (\nu \nabla u) - \nabla \cdot (\mathbf{a}u) + f \\ &= \nabla \cdot (\nu \nabla u) - (\nabla \cdot \mathbf{a})u - \mathbf{a} \cdot \nabla u + f. \end{aligned}$$

This PDE is called the *advection-reaction-diffusion* equation, and it consists of three terms: The first is the *diffusion term* $\nabla \cdot \nu \nabla u$; the second is the *advection term*, that is, $\nabla \cdot (\mathbf{a}u)$, which we expanded in the last equality, and the last is the *reaction term* represented by f. This PDE is in general nonlinear if the reaction term depends on u in a nonlinear fashion.

Remark 1.4. *Often we consider such a problem with multiple substances, for example, u, v, and w. In this case, we obtain a system of PDEs in which the reaction term models the reaction between the various substances, leading to vector-valued reaction functions of the form*

$$\mathbf{f}(u, v, w, x, t) = \begin{cases} f_1(u, v, w, x, t), \\ f_2(u, v, w, x, t), \\ f_3(u, v, w, x, t). \end{cases}$$

If there is no reaction term, the PDE simplifies to

$$u_t = \nabla \cdot \nu \nabla u - \nabla \cdot \mathbf{a}u - \mathbf{a} \cdot \nabla u, \tag{1.10}$$

which is called the advection-diffusion equation.

If the vector field $\mathbf{a}$, *which represents the transport mechanism, is a consequence of thermal convection, then* (1.10) *is called the* convection-diffusion *equation, and in this case* $\mathbf{a}$ *depends in general on* u, *which makes the PDE nonlinear.*

RECHERCHES
SUR LA COURBE QUE FORME UNE CORDE
TENDUË MISE EN VIBRATION,
PAR MR. D'ALEMBERT.

I.

Je me propoſe de faire voir dans ce Memoire, qu'il y a une infinité d'autres courbes que *la Compagne de la Cycloïde allongée*, qui ſatisfont au Probleme dont il s'agit. Je ſuppoſeray toujours 1mo, que les excurſions ou vibrations de la corde ſont fort petites, enſorte que

l'equation generale de la courbe eſt donc
$$y = \psi(t + s) + \Gamma(t - s).$$

Figure 1.18. *Jean-Baptiste le Rond d'Alembert* (16.11.1717–29.10.1783) *and his seminal paper on the modeling of a vibrating string from* 1747.

Finally, if the divergence of the vector field is zero, i.e., $\nabla \cdot \mathbf{a} = 0$, *we obtain for* ν *constant*

$$u_t = \nu \Delta u - \mathbf{a} \cdot \nabla u,$$

a simpler version of the advection-diffusion equation.

1.6 ▪ The Wave Equation

To derive a model for a vibrating violin string, we use *Newton's second law of motion*, which states that mass times acceleration equals the total force acting on the object. If we denote by m the mass, by u the vertical displacement, and by F the forces, then writing the acceleration as u_{tt} leads to

$$m u_{tt} = F.$$

In addition, we know from *Hooke's law* that the restoring force is proportional to the elongation,

$$F \sim u.$$

Combining these two laws, we get an ODE of the form

$$m u_{tt} = -ku,$$

where $k > 0$ is the proportionality constant. This equation is a simplified model for a vibrating string, as it does not take into account the spatial direction.

For a more accurate model in time and space, we follow the derivation of d'Alembert [13]. D'Alembert was an illegitimate child of a chevalier and a writer and grew up in an orphanage. He studied philosophy, law, and arts and also pursued poetry, music, medicine, and mathematics; for a portrait, see Figure 1.18. D'Alembert's idea for understanding the motion of a string was to take a closer look at a small section of the string. On such a section of length Δs, the mass can be expressed as the product of the small interval times the density ρ of the string,

$$m = \rho \Delta s.$$

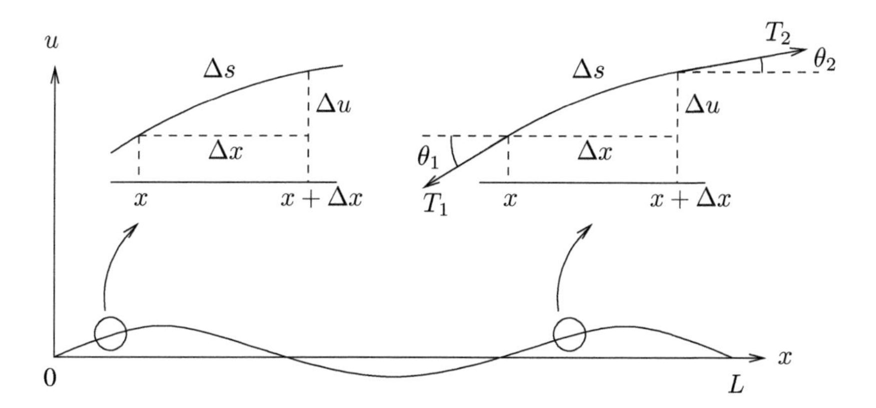

Figure 1.19. *Figure of a vibrating violin string on which we zoomed at two different places to illustrate the decomposition of a small portion of the string in the horizontal and vertical directions and the tensions acting on the string.*

The small section of the string, Δs, can be approximated, as shown in Figure 1.19, by

$$\begin{aligned}\Delta s &\approx \sqrt{\Delta x^2 + \Delta u^2},\\ &= \Delta x\sqrt{1+\left(\frac{\Delta u}{\Delta x}\right)^2},\\ &\approx \Delta x\sqrt{1+(u'(x))^2}.\end{aligned}$$

For the vibration of the string, it is reasonable to assume that the vertical variation $u'(x)$ is small, and thus we neglect this contribution in our model for the mass and use

$$m = \rho\Delta x.$$

The force on the string segment is due to the tension created when one pulls on the string. Let T_1 and T_2 be the tensions in the string at x and $x + \Delta x$, as shown in Figure 1.19. If there is no horizontal movement, then the horizontal components of the tensions must be equal. We name these horizontal components

$$T := T_1\cos(\theta_1) = T_2\cos(\theta_2).$$

The force acting on the string in the vertical direction is given by the difference of the vertical components of the tension at x and $x + \Delta x$,

$$F = T_2\sin(\theta_2) - T_1\sin(\theta_1).$$

Dividing F by T and using both equivalent definitions of T, we obtain

$$\frac{F}{T} = \frac{T_2\sin(\theta_2)}{T_2\cos(\theta_2)} - \frac{T_1\sin(\theta_1)}{T_1\cos(\theta_1)} = \tan(\theta_2) - \tan(\theta_1).$$

But $\tan(\theta_i)$, for $i = 1, 2$, is none other than the slope of the string at the points x and $x + \Delta x$, respectively, which is given by the first spatial derivative of the position $u(x)$. Using this force in *Newton's law of motion* and dividing both sides by $T\Delta x$ gives

$$\begin{aligned}\frac{\rho}{T}u_{tt} &= \frac{1}{\Delta x}(\tan(\theta_2) - \tan(\theta_1)),\\ &= \frac{u_x(x+\Delta x) - u_x(x)}{\Delta x}.\end{aligned}$$

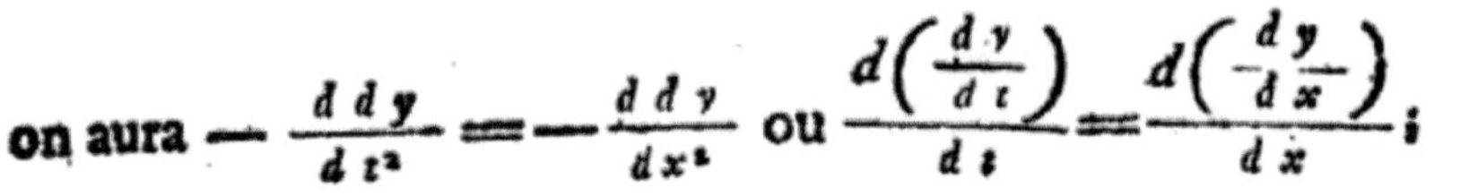

on aura $-\frac{ddy}{dt^2} = -\frac{ddy}{dx^2}$ ou $\frac{d\left(\frac{dy}{dt}\right)}{dt} = \frac{d\left(\frac{dy}{dx}\right)}{dx}$;

Figure 1.20. *The now classical second-order wave equation as it was written by d'Alembert, from* [14]*, which had already been submitted in a first version in* 1755.

Finally, letting Δx tend to zero leads to the so-called *second-order wave equation* in one dimension,

$$u_{tt} = c^2 u_{xx}, \tag{1.11}$$

where $c := \sqrt{T/\rho}$ represents the wave speed. Unlike the heat equation, the wave equation is a hyperbolic PDE, which exhibits oscillatory rather than diffusive behavior; for an example of a solution of the wave equation, see Figure 1.21. We show in Figure 1.20 the wave equation as it appeared in [14], the second of a two-part publication by d'Alembert. A solution formula for this equation already appears in part I of the publication [13] from 1747 (see Figure 1.18 and also Problem 1.9), although d'Alembert did not explicitly formulate the PDE satisfied by it until 1761 in order to address criticisms by Euler and Daniel Bernoulli in 1753 over his previous work.[9] The vibrating string controversy led to important new concepts in the understanding of functions: Instead of being restricted to analytic expressions, the notion now also includes discontinuous functions and functions drawn freehand; see [33].

Remark 1.5.

- *In a domain Ω of dimension two or three,*[10] *with initial and boundary conditions and a source function, the wave equation is*

$$\begin{cases} u_{tt} &= c^2\Delta u + f & \text{in } \Omega \times (0,T), \\ u(\mathbf{x},0) &= u^0(\mathbf{x}) & \text{in } \Omega, \\ u_t(\mathbf{x},0) &= u_t^0(\mathbf{x}) & \text{in } \Omega, \\ u(\mathbf{x},t) &= g(\mathbf{x},t) & \text{on } \partial\Omega, \end{cases}$$

  where $u^0(\mathbf{x})$ is the initial position of the string and $u_t^0(\mathbf{x})$ is its initial velocity. Note that we need two initial conditions because the equation contains a second-order time derivative.

- *An analytic solution of the wave equation can be found using separation of variables, as seen in the case of the heat equation; see Problem* 1.9 *for the case in one dimension.*

- *Using the wave equation, we can approximate the water surface waves created when a platform diver jumps into a pool. The edge of the swimming pool, modeled by homogeneous Neumann conditions, reflects the waves and creates very complicated wave patterns filling the entire swimming pool; see the sequence of graphs in Figure* 1.21.

[9] D'Alembert [14]: "La lecture de leurs Mémoires & les miens suffiroit peut-être pour me mettre à couvert de leurs attaques; car chacun de ces grands Géometres, pris séparément, semble m'accorder ce que l'autre me nie."

[10] The three-dimensional case was discovered by Euler in 1766.

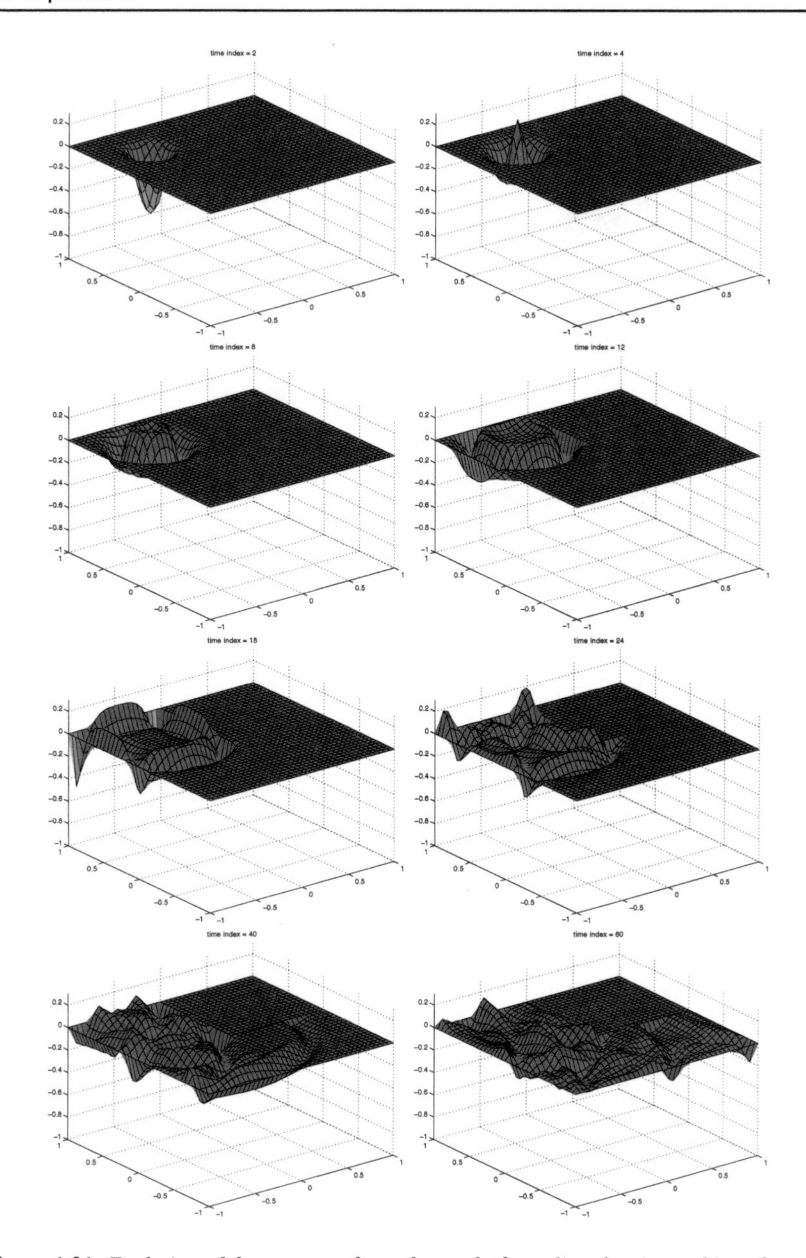

Figure 1.21. *Evolution of the water surface after a platform diver has jumped into the pool.*

1.7 ▪ Maxwell's Equations

In our current environment of mobile phones and wireless computer networks, electric and magnetic fields are omnipresent. The complete equations satisfied by these fields were first formulated and studied by Maxwell in [6], with an earlier version appearing already in 1856; see Figure 1.22. Maxwell's achievement was to reduce everything that was known about electricity and magnetism at his time to a system of 20 equations in 20 unknowns and his introduction of the notion of an electromagnetic field, in contrast to the force lines used by Faraday. In modern notation, one uses only six unknowns:

[From the *Transactions of the Cambridge Philosophical Society*, Vol. x. Part i.]

VIII. *On Faraday's Lines of Force.*

[Read *Dec.* 10, 1855, and *Feb.* 11, 1856.]

The present state of electrical science seems peculiarly unfavourable to speculation. The laws of the distribution of electricity on the surface of conductors have been analytically deduced from experiment; some parts of the mathematical theory of magnetism are established, while in other parts the experimental data are wanting; the theory of the conduction of galvanism and that of the mutual attraction of conductors have been reduced to mathematical formulæ, but have not fallen into relation with the other parts of the science. No electrical theory can now be put forth, unless it shews the connexion not only between electricity at rest and current electricity, but between the attractions and inductive effects of electricity in both states. Such a theory must accurately satisfy those laws, the mathematical form of which is known, and must afford the means of calculating the effects in the limiting cases where the known formulæ are inapplicable.

Figure 1.22. *James Clerk Maxwell* (13.06.1831–05.11.1879)*, and the beginning of his seminal paper "On Faraday's lines of force."*

We denote the time-dependent electric vector field by

$$\mathbf{E} : \mathbb{R}^4 \longrightarrow \mathbb{R}^3$$

and the time-dependent magnetic field by

$$\mathbf{H} : \mathbb{R}^4 \longrightarrow \mathbb{R}^3.$$

The PDEs governing these fields, now known as Maxwell's equations, consist of two physical laws:

1. *The Maxwell–Ampere law*:

$$-\varepsilon \frac{\partial \mathbf{E}}{\partial t} + \nabla \times \mathbf{H} - \mathbf{J} = 0. \tag{1.12}$$

   Here ε is the *electric permittivity* and $\mathbf{J}$ is the *current density*. The Maxwell–Ampere law relates the current to the change in the electric field and the rotation of the magnetic field.[11]

2. *The Maxwell–Faraday law*:

$$\mu \frac{\partial \mathbf{H}}{\partial t} + \nabla \times \mathbf{E} = 0. \tag{1.13}$$

   Here μ is the *magnetic permeability*. The Maxwell–Faraday law relates the change of the magnetic field to the rotation of the electric field. There are no physical magnetic currents, and there is no magnetic damping.

[11]It is tempting to use *Ohm's law*, $\mathbf{J} = \sigma \mathbf{E}$, where $\sigma > 0$ is the *electric conductivity*, to replace $\mathbf{J}$ in (1.12) and thus obtain a system with damping to the electric field. Ohm's law in this form is, however, very different in character from Maxwell's equations; see, for example, [24, p. 289], where the author says that "it's not really a true law, in the sense of Gauss's law or Ampère's law; rather, it is a 'rule of thumb' that applies pretty well to many substances. You're not going to win a Nobel Prize for finding an exception. In fact, when you stop to think about it, it's a little surprising that Ohm's law ever holds," and [46, p. 129], where the author says, "It is an empirical law, a generalization derived from experiment, not a theorem that must be universally obeyed. In fact, Ohm's law is bound to fail in the case of any particular material if the electric field is too strong"; see also the recent publication specifically about this topic [15], where the author claims that the two fields $\mathbf{E}$ are really different quantities: "It is found that, in general, the electric field in Ohm's law cannot be identified with the electric field as defined by Maxwell's equations." The correct approach to include Ohm's law is to substitute $\mathbf{J} = \sigma \mathbf{E} + \mathbf{J}_e$ into Maxwell's equations, which then include the current source due to the electric field in the domain and a possible further external current source $\mathbf{J}_e$.

Both (1.12) and (1.13) together form a system of six linear differential equations for the six vector field components in $\mathbf{E}$ and $\mathbf{H}$, and this system is called Maxwell's equations.

The electric field $\mathbf{E}$ and the magnetic field $\mathbf{H}$ must satisfy in addition the divergence conditions

$$\nabla \cdot (\varepsilon \mathbf{E}) = \rho \; (\textit{Gauss's law}), \quad \nabla \cdot (\mu \mathbf{H}) = 0 \; (\textit{Gauss's law for magnetism}). \quad (1.14)$$

However, when the material parameters ε, μ, and σ are independent of time, these conditions are in fact redundant, as long as they are satisfied by $\mathbf{E}$ and $\mathbf{H}$ at initial time. To show this, first note that the divergence of the curl vanishes, as one can see by a direct calculation or using the Maple command

```
diverge(curl([E1(x,y,z),E2(x,y,z),E3(x,y,z)],[x,y,z]),[x,y,z]);
```

from the `linalg` package. Applying the divergence operator to (1.13), we get

$$\nabla \cdot \left(\mu \frac{\partial \mathbf{H}}{\partial t} \right) = \frac{\partial}{\partial t} (\nabla \cdot (\mu \mathbf{H})) = 0.$$

Thus, $\nabla \cdot (\mu \mathbf{H})$ is constant in time, so if it vanishes at the initial time $t = 0$, it will continue to vanish for all time t. Similarly, by taking the divergence of (1.12),

$$\nabla \cdot \left(\varepsilon \frac{\partial \mathbf{E}}{\partial t} + \mathbf{J} \right) = \frac{\partial}{\partial t} (\nabla \cdot (\varepsilon \mathbf{E})) + \nabla \cdot \mathbf{J} = 0,$$

and then using the conservation of the electric charge,

$$\frac{\partial \rho}{\partial t} + \nabla \cdot \mathbf{J} = 0,$$

we obtain

$$\frac{\partial}{\partial t} (\nabla \cdot (\varepsilon \mathbf{E}) - \rho) = 0,$$

which implies that if Gauss's law $\nabla \cdot (\varepsilon \mathbf{E}) = \rho$ is verified at initial time, then it is also verified for all time.

1.8 ▪ Navier–Stokes Equations

The vector field $\mathbf{u} : \mathbb{R}^4 \to \mathbb{R}^3$ of the velocity of an incompressible Newtonian fluid satisfies the *Navier–Stokes equations*, discovered by Claude-Louis Navier and George Gabriel Stokes (see Figure 1.23):

$$\left\{ \begin{array}{rcl} \mathbf{u}_t + \mathbf{u} \cdot \nabla \mathbf{u} + \nabla p & = & \nu \Delta \mathbf{u}, \\ \nabla \cdot \mathbf{u} & = & 0. \end{array} \right.$$

Here, the scalar quantity $p : \mathbb{R}^4 \to \mathbb{R}$ is the pressure and ν is the viscosity. We recognize in the first equation the three familiar terms of an advection-diffusion equation, which transports and diffuses the entire velocity field $\mathbf{u}$. But now the transport field is $\mathbf{u}$ itself, which makes this equation nonlinear. In addition, there is the term ∇p, which is necessary to balance the requirement of the second equation, which states that the velocity field $\mathbf{u}$ must be divergence free, and thus the fluid is incompressible.

Depending on the concrete situation, various simplifications of the incompressible Navier–Stokes equations are useful.

Figure 1.23. *Claude-Louis Navier* (10.02.1785–21.08.1836), *and George Gabriel Stokes* (13.08.1819–01.02.1903).

Remark 1.6.

- *Neglecting the nonlinear term, we obtain the Stokes equations,*

$$\begin{cases} \mathbf{u}_t + \nabla p &= \nu\Delta\mathbf{u}, \\ \nabla \cdot \mathbf{u} &= 0, \end{cases}$$

  which are also often used in the stationary case, i.e., when $\mathbf{u}_t$ *is zero.*

- *In a so-called perfect fluid, we can neglect the viscosity* ν *and obtain the simplified system*

$$\begin{cases} \mathbf{u}_t + \mathbf{u} \cdot \nabla\mathbf{u} + \nabla p &= 0, \\ \nabla \cdot \mathbf{u} &= 0. \end{cases}$$

  These equations are called the incompressible Euler equations, *as they appeared first in published form in* [37]; *see Figure* 1.24.

- *One can also linearize the system around a given velocity* $\mathbf{u}_0(\mathbf{x})$, *which leads, when also neglecting the viscosity, to the* Oseen equations,

$$\begin{cases} \mathbf{u}_t + \mathbf{u}_0 \cdot \nabla\mathbf{u} + \nabla p &= 0, \\ \nabla \cdot \mathbf{u} &= 0. \end{cases}$$

An excellent reference for physical modeling using PDEs is the chapter "Les modèles physiques" in the first volume of *Analyse mathématique et calcul numérique pour les sciences et les techniques* by Robert Dautray and Jacques-Louis Lions [10]. This classical volume has also been translated into English; see [11].

1.9 ▪ Elliptic Problems

As we have seen in the last few sections, many problems of physical interest are initial-boundary value problems, in which we seek a time-dependent solution $u(\mathbf{x}, t)$ starting with some initial conditions $u(\mathbf{x}, 0)$ and boundary conditions $u|_{\partial\Omega}$. An equally interesting class of problems are *elliptic problems*, in which we seek a solution $u(\mathbf{x})$ defined

274

PRINCIPES GÉNÉRAUX
DU MOUVEMENT DES FLUIDES.
PAR M. EULER.

I.

Avant établi dans mon Mémoire précedent les principes de l'équilibre des fluides le plus généralement, tant à l'égard de la diverſe qualité des fluides, que des forces qui y puiſſent agir ; je me propoſe de traiter ſur le même pied lè mouvement des fluides, & de rechercher les principes géneraux, ſur lesquels toute la ſcience du mouvement des fluides eſt fondée. On comprend aiſément que cette matiere eſt beaucoup plus difficile, & qu'elle renferme des recherches incomparablement plus profondes : cependant j'eſpère d'en venir auſſi heureuſement à bout, de ſorte que s'il y reſte des difficultés, ce ne ſera pas du côté du méchanique, mais uniquement du côté de l'analytique: cette ſcience n'étant pas encore portée à ce degré de perfection, qui feroit néceſſaire pour développer les formules analytiques, qui renferment les principes du mouvement des fluides.

XXI. Nous n'avons donc qu'à égaler ces forces accélératrices avec les accélerations actuelles que nous venons de trouver, & nous obtiendrons les trois équations ſuivantes :

$$P - \frac{1}{q}\left(\frac{dp}{dx}\right) = \left(\frac{du}{dt}\right) + u\left(\frac{du}{dx}\right) + v\left(\frac{du}{dy}\right) + w\left(\frac{du}{dz}\right)$$

$$Q - \frac{1}{q}\left(\frac{dp}{dy}\right) = \left(\frac{dv}{dt}\right) + u\left(\frac{dv}{dx}\right) + v\left(\frac{dv}{dy}\right) + w\left(\frac{dv}{dz}\right)$$

$$R - \frac{1}{q}\left(\frac{dp}{dz}\right) = \left(\frac{dw}{dt}\right) + u\left(\frac{dw}{dx}\right) + v\left(\frac{dw}{dy}\right) + w\left(\frac{dw}{dz}\right)$$

Si nous ajoutons à ces trois équations premièrement celle, que nous a fournie la conſidération de la continuité du fluide :

$$\left(\frac{dq}{dt}\right) + \left(\frac{d.qu}{dx}\right) + \left(\frac{d.qv}{dy}\right) + \left(\frac{d.qw}{dz}\right) = 0,$$

Figure 1.24. *Leonhard Euler* (15.04.1707–18.09.1783) *on the Swiss* 10-*franc note and the invention of Euler's equations in* 1757.

on the spatial domain Ω only, given some boundary conditions $u|_{\partial\Omega}$. A general second-order linear elliptic equation in d dimensions is of the form $Lu = f$, where

$$Lu = -\sum_{i=1}^{d}\sum_{j=1}^{d} \frac{\partial}{\partial x_i}\left(a_{ij}(\mathbf{x})\frac{\partial u}{\partial x_j}\right) + \sum_{i=1}^{d} b_i(\mathbf{x})\frac{\partial u}{\partial x_i} + c(\mathbf{x})u,$$

with the additional (uniform) ellipticity condition

$$\sum_{i=1}^{d}\sum_{j=1}^{d} a_{ij}(\mathbf{x})\xi_i\xi_j \geq \theta\|\boldsymbol{\xi}\|^2 \qquad \forall\boldsymbol{\xi} \in \mathbb{R}^d$$

for some $\theta > 0$. In other words, we require the symmetric tensor $A(\mathbf{x}) := (a_{ij}(\mathbf{x}))_{i,j=1}^d$ to be positive definite, with minimum eigenvalue at least θ almost everywhere. Elliptic problems often arise from the following considerations.

Semidiscretization in time: One way of solving a time-dependent problem is to subdivide the time interval $[0, T]$ into a time grid $0 = t^0 < t^1 < \cdots < t^N = T$ and approximate the time derivative by a *finite difference in time*. The simplest version is the first-order finite difference

$$\frac{\partial u}{\partial t}(\mathbf{x}, t^n) = \frac{u(\mathbf{x}, t^{n+1}) - u(\mathbf{x}, t^n)}{\Delta t^n} + O(\Delta t^n),$$

where $\Delta t^n := t^{n+1} - t^n$ is the nth time step. For the heat equation

$$\frac{\partial u}{\partial t} = \Delta u + f,$$

replacing the time derivative with the finite difference approximation leads to

$$\frac{u^{n+1}-u^n}{\Delta t^n} = \Delta u + f,$$

where $u^n \approx u(\cdot, t^n)$ is an approximation of the solution u at time t^n. Note that we have not specified the time t at which u and f on the right-hand side are evaluated. If these terms are taken at t^n, we obtain the *explicit method*

$$u^{n+1} = u^n + \Delta t^n(\Delta u^n + f(\cdot, t^n)),$$

which is called *forward Euler* and does not require solving a boundary value problem. However, for stability reasons, one often prefers to evaluate u and f at time t^{n+1} instead, leading to the *implicit method*

$$-\Delta u^{n+1} + \frac{1}{\Delta t^n}u^{n+1} = \frac{1}{\Delta t^n}u^n + f(\cdot, t^{n+1}),$$

which is called *backward Euler*. If we define $g^n := u^n/\Delta t^n + f(\cdot, t^{n+1})$, we obtain an elliptic problem in u^{n+1} of the form

$$-\Delta u^{n+1} + \alpha u^{n+1} = g^n, \tag{1.15}$$

where $\alpha_n = 1/\Delta t^n > 0$ is a positive parameter that depends on the time step size. Thus, to solve the time-dependent problem numerically, one would solve a sequence of elliptic problems (1.15) for $n = 0, 1, \ldots, N-1$. Other time discretizations are possible (e.g., the Crank–Nicolson method), but any implicit method will invariably produce a sequence of elliptic problems that must be solved at each time step.

Steady-state/equilibrium solutions: A second way in which elliptic problems arise is through *steady-state solutions*. As we have seen in section 1.4, the evolution of temperature in a spatial domain Ω is described by the *heat equation*

$$\frac{\partial u}{\partial t} - \nabla \cdot (a(\mathbf{x})\nabla u) = f(\mathbf{x}, t),$$

where $a(\mathbf{x}) \geq a_{\min} > 0$. If neither forcing term f nor the boundary conditions $u|_{\partial\Omega}$ depend on time, then it can be shown that $\frac{\partial u}{\partial t} \to 0$ and that $u(\mathbf{x}, t) \to \bar{u}(\mathbf{x})$, where $\bar{u}(\mathbf{x})$ satisfies the elliptic equation

$$-\nabla \cdot (a(\mathbf{x})\nabla \bar{u}) = f(\mathbf{x}). \tag{1.16}$$

If $a(\mathbf{x}) = \nu > 0$ is constant in space, (1.16) is known as the *Poisson equation*; see also Example 1.11. If, in addition, we have $f \equiv 0$ and $a(\mathbf{x}) = \nu = 1$, then it is called the *Laplace equation*. Note that the Poisson equation also describes the steady-state solution of the wave equation

$$u_{tt} = c^2\Delta u + f.$$

Another situation where the Poisson equation appears is in electrostatics. Consider once again the *Maxwell–Faraday law* (1.13)

$$\mu\frac{\partial \mathbf{H}}{\partial t} + \nabla \times \mathbf{E} = 0.$$

If the system is at steady state such that $\frac{\partial \mathbf{H}}{\partial t} = 0$, then the electric field is *curl-free*; i.e., we have $\nabla \times \mathbf{E} = 0$. This implies $\mathbf{E}$ is the gradient of a scalar field ψ, i.e., $\mathbf{E} = \nabla\psi$. Substituting this into Gauss's law (1.14) gives

$$\nabla \cdot (\epsilon\nabla\psi) = \rho.$$

If the electric permittivity $\epsilon > 0$ is constant in space, we obtain once again the Poisson equation $\Delta\psi = \rho/\epsilon$.

For the advection-diffusion equation

$$u_t = \nabla \cdot (\nu \nabla u) - \nabla \cdot (\mathbf{a}u) + f,$$

the steady-state solution is given by

$$-\nabla \cdot (\nu \nabla u) + \nabla \cdot (\mathbf{a}u) = f,$$

which is also an elliptic problem.

The steady-state solution of a Navier–Stokes equation satisfies

$$\begin{cases} \mathbf{u} \cdot \nabla \mathbf{u} + \nabla p = \nu \Delta \mathbf{u}, \\ \nabla \cdot \mathbf{u} = 0. \end{cases}$$

This is a nonlinear equation in $\mathbf{u}$. For small velocities (relative to the viscosity ν), one can neglect the quadratic term (cf. Remark 1.6) to obtain the *steady-state Stokes equation*

$$\begin{cases} -\nu \Delta \mathbf{u} + \nabla p = 0, \\ \nabla \cdot \mathbf{u} = 0. \end{cases} \tag{1.17}$$

Although the first equation in (1.17) resembles a Poisson equation, the fact that it is coupled with the divergence-free constraint makes the problem a saddle point problem with very different properties that require different discretization techniques than the standard elliptic Poisson equation.

Time-harmonic regimes: A third situation in which one encounters elliptic equations is in the so-called *time-harmonic regime*, in which the forcing term and boundary conditions are periodic with frequency ω. In the case of the heat equation with homogeneous boundary conditions, we assume that the forcing term satisfies $f(\mathbf{x}, t) = \tilde{f}(\mathbf{x})e^{i\omega t}$, so that the solution approaches the time-periodic solution $u(\mathbf{x}, t) \to \tilde{u}(\mathbf{x})e^{i\omega t}$ as $t \to \infty$, where $\tilde{u}(\mathbf{x})$ satisfies

$$i\omega\tilde{u} - \Delta\tilde{u} = \tilde{f}.$$

This is the *shifted Laplace equation* with an imaginary shift $i\omega$. Similarly, to look for a time-harmonic solution of the second-order wave equation, we suppose that the solution is of the form

$$u(\mathbf{x}, t) = \tilde{u}(\mathbf{x})e^{i\omega t},$$

where ω is a given frequency. If we introduce this ansatz into the wave equation, we get

$$-\omega^2\tilde{u}(\mathbf{x})e^{i\omega t} = \Delta\tilde{u}e^{i\omega t} + f.$$

Now simplifying the exponential term, collecting terms depending on $\tilde{u}$ on the left, and defining $\tilde{f} := fe^{-i\omega t}$, we obtain

$$-(\Delta + \omega^2)\tilde{u} = \tilde{f},$$

which is called the *Helmholtz equation*.[12] Note that if we change the sign inside the parentheses, we find a diffusion equation of the same form as (1.15). Thus, the sign is

[12]This equation was studied by Helmholtz [30] to obtain a better physical understanding of organ pipes; see also [21].

of paramount importance for the qualitative behavior of the solutions: The Helmholtz equation comes from a wave equation, and solutions are oscillatory, whereas with the sign changed, solutions represent a diffusive phenomenon, which rapidly damps out all oscillatory components, as we have seen in the one-dimensional example earlier.

Because elliptic problems occur naturally as subproblems in many time-dependent PDEs and are interesting in their own right, a substantial amount of research has been devoted to developing effective numerical methods for these problems. In the next few chapters, we will introduce four major classes of numerical methods for elliptic problems, each based on a different principle:

- Finite difference methods, based on Taylor expansion;
- finite volume methods, based on discrete conservation;
- spectral methods, based on Fourier analysis;
- finite element methods, based on the variational principle.

1.10 ▪ Problems

Problem 1.1 (Maple). For all Maple commands, the online help can be accessed using the question mark, for example, `? series`.

(i) Use the commands `taylor` and `series` to compute expansions of the function $\sin(x)$, $\cos(\sqrt{x})^2$, and $\frac{1}{x^2+x}$ about 0. What is the difference between the two commands?

(ii) For the function $\sin(x)$, draw a graph of the function and its Taylor series, truncated after 2, 4, and 6 terms. Use the command `convert` to convert the Taylor series into a polynomial and, if you need to, the command `display` to display several graphs in one drawing.

(iii) Use the commands `sum` and `product` to compute the sum of the first n integers cubed and the product of the first n even integers.

(iv) Use the commands `solve` and `allvalues` to find the solutions of $x^2+2x-x^5 = 0$ and $x^4+x+1=0$.

(v) We consider the expression

```
f:=exp(-(x^2+y^2))
```

and the function

```
g:=(x,y)->exp(-(x^2+y^2)).
```

Compute the derivatives of `f` and `g` using the command `diff` and the operator `D`. Explain the difference between these two commands. Draw `f` and `g` on the domain $[-2,2]\times[-2,2]$ with the command `plot3d`.

Problem 1.2 (MATLAB). For all MATLAB commands, one can access the online help using the command `help` followed by the name of the command, for example, `help plot`.

(i) Use MATLAB to solve the system of linear equations $Ax = b$ for the following:

1. A, a randomly generated 3×3 matrix, and b, a randomly generated column vector with 3 entries (use the command `rand`).

2. $A = \begin{pmatrix} -2 & 1 & 0 \\ 1 & -2 & 1 \\ 0 & 1 & -2 \end{pmatrix}$ and $b = \begin{pmatrix} 1 \\ 2 \\ 3 \end{pmatrix}$.

(ii) Use the graphics capabilities of MATLAB to plot the following graphs:

1. The graph of the functions f, g, and h between 0 and 5 on the same plot, using the command `plot`:

$$f(x) = \exp\left(\frac{1}{x+1}\right),$$
$$g(x) = e\left(1 - x + \frac{3}{2}x^2\right),$$
$$h(x) = e\left(1 - x + \frac{3}{2}x^2 - \frac{13}{6}x^3 + \frac{73}{24}x^4\right),$$

where e is Euler's constant.

2. The graph of the following functions, using the commands `semilogy`, `loglog`, and `plot`:

$$f(x) = \exp\left(\frac{1}{x+1}\right),$$
$$g(x) = \frac{1}{x+1},$$
$$h(x) = \ln(x).$$

3. Represent the surface given by the function $f(x, y) = \sin(x)y + \frac{1}{1+y^2}$ using the commands `mesh`, `surf`, and `contour`. The command `meshgrid` is very useful here.

Problem 1.3 (pendulum problem). Solve the pendulum equation (1.1) numerically with Maple. Use as initial conditions $\theta(0) = \frac{1}{10}$ and $\dot{\theta}(0) = 0$, as in the numerical example with MATLAB in Section 1.2. Try both the `series` and the `numeric` options in the `dsolve` command of Maple (type `'? dsolve'` in Maple to get the online help). Plot the two results, and compare them with the results obtained with MATLAB. What do you observe? (Hint: the output argument `output=listprocedure` is useful for the `numeric` option.)

Problem 1.4 (pendulum equation). The goal of this problem is to solve numerically the pendulum equation using MATLAB and to compare the result to the analytic solution of the equation that uses the small angle approximation. The pendulum equation is given by

$$\ddot{\theta} = -\frac{g}{L}\sin(\theta), \tag{1.18}$$

and the small angle approximation leads to the equation

$$\ddot{\theta} = -\frac{g}{L}\theta. \tag{1.19}$$

1. Use the ODE solver `ode45` from MATLAB to solve (1.18), and compare your solution to the exact solution of (1.19) for various initial conditions. Change both the initial position and velocity.

2. Up to which angle is the small angle approximation still good, assuming that the initial velocity is zero?

3. What is qualitatively the behavior of the pendulum if the initial velocity is very large? How does this translate in the numerical solution computed in MATLAB?

Problem 1.5 (modeling). A tank contains two kilograms of salt dissolved in 400 liters of water. We pour continuously, at a rate of 10 liters per minute, a salty solution with concentration of 0.3 kg/l into the tank. We suppose that the two liquids get instantaneously well mixed, with the help of a stirrer, and that the volume of the tank is constant, which means that there is an overflow where the mixed liquid can leave the tank.

1. Find a mathematical model for the process.

2. Find the quantity of salt in the tank after 10 minutes.

3. What concentration will one approach if one waits for a very long time?

4. What happens if the tank has a volume of 1000 litres; i.e., at the beginning the salty solution does not overflow?

5. What concentration would one approach if the tank had an infinite volume, if one waits for a very long time?

Problem 1.6 (Newton's law of cooling). If an object with temperature T is put into a surrounding environment of constant temperature M, the temperature variation of the object is proportional to the difference of the temperatures $M - T$. This leads to the differential equation

$$\frac{dT}{dt} = k(M - T),$$

where k is the constant of proportionality.

(i) Solve the differential equation analytically for $T(t)$.

(ii) A student buys a thermometer to check his body temperature. Feeling healthy, he checks his temperature: Before starting, the thermometer shows $22°C$ for the surrounding temperature, and after 10 minutes under the arm, as indicated on the package, the thermometer shows 37.38°C.

A few days later, the student feels unwell and measures his temperature again. This time, the thermometer shows $19°C$ for the surrounding temperature, and, being impatient, the student already checks after having the thermometer only 1.5 minutes under the arm to find the temperature 30°C.

1. Compute the temperature if the thermometer had been left 10 minutes, as required.
2. Compute the exact temperature of the student. Does he have a fever?
3. Does the student really need to wait 10 minutes, as recommended on the package?

Problem 1.7 (Lotka–Volterra). Try finding a closed-form solution of the system of Lotka–Volterra,

$$\dot{x} = ax - bxy, \tag{1.20}$$

$$\dot{y} = -cy + dxy, \tag{1.21}$$

which models the interaction of a predator population $y(t)$ with a prey population $x(t)$, using Maple. Choose for the constants $a = b = c = d = 1$. Plot the solution, and compare it to the numerical solution in Maple and MATLAB. What is the long-term behavior of the two populations?

Problem 1.8 (divergence theorem). We consider a bounded open domain $\Omega \subset \mathbb{R}^n$ with boundary $\partial\Omega$. We denote by $\mathbf{x}$ a point $(x_1, \dots, x_n)$ in $\mathbb{R}^n$.

Definition 1.12. *The boundary $\partial\Omega$ is said to belong to the class $\mathcal{C}^k$, $k \in \mathbb{Z}$, if for each $\mathbf{x}_0 \in \partial\Omega$, there exist $r > 0$, an $n \times n$ rotation matrix Q (i.e., $Q^T Q = I$ and $\det(Q) = 1$), and a function $\gamma : \mathbb{R}^{n-1} \to \mathbb{R}$ in $\mathcal{C}^k$ such that*

$$\Omega \cap B(\mathbf{x}_0, r) = \{\mathbf{x} \in B(\mathbf{x}_0, r) : \mathbf{x} = \mathbf{x}_0 + Q\mathbf{y},\ y_n > \gamma(y_1, \dots, y_{n-1})\}.$$

Remark. *If $\partial\Omega$ is $\mathcal{C}^1$, then the unit outer normal vector $\mathbf{n}$ is well defined. In this case, we define the normal derivative of a function $u : \bar{\Omega} \to \mathbb{R}$ in $\mathcal{C}^1$ by*

$$\frac{\partial u}{\partial n} = \nabla u \cdot \mathbf{n}.$$

Theorem 1.13 (Gauss–Green). *If $\partial\Omega \in \mathcal{C}^1$ and $u \in \mathcal{C}^1(\bar{\Omega})$, then*

$$\int_\Omega u_{x_i} d\mathbf{x} = \int_{\partial\Omega} u n_i ds, \quad i = 1, 2, \dots, n.$$

We first suppose, after a suitable rotation, that the domain is defined by two functions f and g, as shown in Figure 1.25(a). The domain Ω is thus given by

$$\Omega = \{(x, y) \in \Omega : a < x < b \text{ and } g(x) < y < f(x)\}.$$

Complete the following outline of the proof of the Gauss–Green theorem:

1. Define $\partial\Omega$ as the union of two level sets defined by f and g.
2. Give the unit outer normal vectors on the boundary $\partial\Omega$.

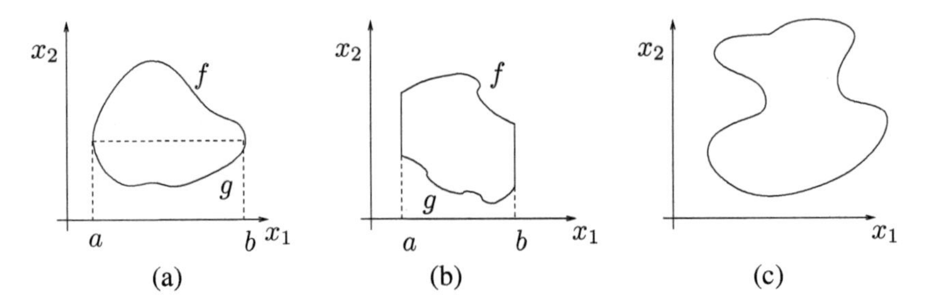

Figure 1.25. *Three steps in the proof of the divergence theorem.*

3. Compare the values of $\int_{\partial\Omega} un_2 ds$ and $\int_\Omega u_{x_2} d\mathbf{x}$ using the infinitesimal relations
$$ds^2 = dx^2 + df^2 \quad \text{and} \quad ds^2 = dx^2 + dg^2.$$
4. Is the Gauss–Green theorem still valid for a domain like in Figure 1.25(b)?
5. What can be done if the domain is like in (c) of Figure 1.25?
6. Use the Gauss–Green theorem to prove the following result:

Theorem 1.14 (divergence theorem). *If $\partial\Omega$ is in C^1 and $\mathbf{f} : \bar{\Omega} \to \mathbb{R}^n$ is a vector field in C^1, then*
$$\int_\Omega \nabla \cdot \mathbf{f} d\mathbf{x} = \int_{\partial\Omega} \mathbf{f} \cdot \mathbf{n} ds.$$

Problem 1.9 (wave equation). We study the wave equation
$$u_{tt} = c^2 u_{xx}.$$

1. (D'Alembert's method). Use the factorization of the wave operator $\partial_{tt} - c^2\partial_{xx}$ into $(\partial_t + c\partial_x)(\partial_t - c\partial_x)$ to show that the solution u of the wave equation can be written in the form $u(x,t) = f(ct + x) + g(ct - x)$.

   Interpret this solution for a string of infinite length, with initial condition $u(x,0) = e^{-x^2}$ and $u_t(x,0) = 0$, and provide a drawing explaining your interpretation.
2. Use the method of separation of variables, setting $u(x,t) = a(x)b(t)$, in order to find a solution of the wave equation in the interval $(0, L)$ with homogeneous boundary conditions.
3. What are the fundamental differences with the solution of the heat equation seen in subsection 1.4?
4. Solve the equation with initial conditions $u(x,0) = \sin(x)$, $u_t(x,0) = 0$, and wave speed $c = 1$ on the interval $(0,\pi)$ and homogeneous boundary conditions.

Problem 1.10 (conservation laws). Consider the flow inside a thin tube of a fluid carrying tiny particles called *tracers*. We denote by $q(x,t)$ the density of these tracers, and we model the thin tube by a one-dimensional line.

1. Give a mathematical expression for the total mass of the tracers in the tube between two points x_1 and x_2.

2. The flow of the fluid describes the flux of the tracers, and we denote this flux by f. Find a conservation law which models the variation of the total mass in the section of the tube between x_1 and x_2.

3. Give an expression of the flux as a function of the velocity $v(x, t)$ of the fluid.

4. Derive a PDE using the previously obtained results.

5. Suppose now that the fluid speed is constant in the tube, i.e., $v(x, t) = c$. Verify that $g(x - ct)$ is a solution of the PDE you found for any sufficiently smooth function g.

6. If we set $X(t) := a + ct$, where a is a constant, show that $q(X(t), t)$ is independent of time. What can one thus say about the solutions of the PDE you found?

Chapter 2
The Finite Difference Method

Das Verfahren besteht nun darin, dass wir in den Querschnitt ein Netz von Quadraten einzeichnen und die Differentialgleichung durch eine Differenzengleichung ersetzen, in der die Differenzen aufeinanderfolgender Werte von x und y gleich der Seite der quadratischen Maschen des Netzes sind.[a]

Carl Runge, Über eine Methode die partielle Differentialgleichung Δu = Constans numerisch zu integrieren, 1908

Ersetzt man bei den klassischen linearen Differentialgleichungsproblemen der mathematischen Physik die Differentialquotienten durch Differenzenquotienten in einem—etwa rechtwiklig angenommenen—Gitter, so gelangt man zu algebraischen Problemen von sehr durchsichtiger Struktur. Die vorliegende Arbeit untersucht nach einer elementaren Diskussion dieser algebraischen Probleme vor allem die Frage, wie sich die Lösungen verhalten, wenn man die Maschen des Gitters gegen Null streben lässt.[b]

R. Courant, K. Friedrichs, H. Levy, Über die partiellen Differenzengleichungen der Mathematischen Physik, 1928

Die vorliegende Arbeit gibt eine genaue Abschätzung der Fehler, welche bei der Lösung der linearen partiellen Differenzengleichung von elliptischem Typus mittels des Differenzenverfahrens entstehen.[c]

S. Gershgorin, Fehlerabschätzung für das Differenzenverfahren zur Lösung partieller Differentialgleichungen, 1930

[a] The procedure consists of drawing a network of squares and replacing the differential equation by a difference equation in which successive values of x and y equal the side of the square mesh of the network.

[b] If one replaces in the classical linear differential equations of mathematical physics the differential quotients by difference quotients in a rectangular mesh, one arrives at algebraic problems of very transparent structure. Here we investigate, after an elementary discussion of these algebraic problems, mainly the question of how their solutions behave when the mesh size goes to zero.

[c] This chapter contains a precise estimate of the error which arises when one solves elliptic PDEs approximately using a finite difference method.

The *finite difference method* is an easy-to-understand method for obtaining approximate solutions of PDEs. The method was introduced by Runge in 1908 to understand the torsion in a beam of arbitrary cross section, which results in having to solve a Poisson equation; see the quote above and also Figure 2.1. The finite difference method is based on an approximation of the differential operators in the equation by finite differences,

Über eine Methode die partielle Differentialgleichung
$\Delta u =$ Constans numerisch zu integrieren.

Von C. Runge in Göttingen.

Die Bestimmung der Torsion eines zylindrischen Stabes mit beliebigem Querschnitt läuft darauf hinaus, eine Funktion u der rechtwinkligen Koordinaten x, y zu finden, die im Innern des Querschnitts der partiellen Differentialgleichung

$$\frac{\partial^2 u}{\partial x^2} + \frac{\partial^2 u}{\partial y^2} = \text{Const.}$$

genügt, während sie auf dem Rande des Querschnittes Null ist.[1]) Ich möchte hier eine Methode entwickeln, um diese Aufgabe zu lösen und zwar will ich sie bei einem speziellen Beispiele durchführen, weil es gerade auf die numerische Ausführung ankommt. Aus dem speziellen Beispiel ergibt sich von selbst, wie man in andern Fällen zu verfahren hat.

Fig. 1.

Figure 2.1. *Carl David Tolmé Runge* (30.08.1856–03.01.1927), *and the beginning of the manuscript where he introduced the finite difference method for the Poisson equation* [52].

which is natural since the derivatives themselves are defined to be the limit of a finite difference,

$$f'(x) := \lim_{h \to 0} \frac{f(x+h) - f(x)}{h}.$$

The method was immediately put to use by Richardson, who tried a retroactive forecast of the weather for May 20, 1910, by direct computation. The forecast failed dramatically because of roughness in the initial data that led to unphysical surges in pressure, but the method was essentially correct and led to Richardson's famous book on numerical weather prediction [48] in 1922. A first convergence proof for the finite difference method was given in 1928 in the seminal paper by Courant, Friedrichs, and Lewy [9], and the first error estimate is due to Gershgorin [23]; see the quotes above and also the excellent historical review [59]. We will start by looking at the Poisson equation to explain the finite difference method.

2.1 ▪ Finite Differences for the Two-Dimensional Poisson Equation

We want to compute an approximate solution of the Poisson equation

$$\left\{ \begin{array}{rcll} \Delta u & = & f & \text{in } \Omega, \\ u & = & g & \text{on } \partial\Omega. \end{array} \right. \tag{2.1}$$

We first assume that the domain Ω is simply the unit square. By prescribing the function value on the boundary $\partial\Omega$ as in (2.1), we are said to be imposing a *Dirichlet boundary condition*. Later, we will consider the case where we prescribe a given flux across the boundary, so that $\frac{\partial u}{\partial n} := \nabla u \cdot \mathbf{n} = g$, where $\mathbf{n}$ is the unit outer normal vector to the boundary $\partial\Omega$; this is known as a *Neumann boundary condition*.

The idea of the finite difference method is to approximate the derivatives in the PDE using a truncated Taylor series in each variable. For example, after perturbing the x variable by a distance h, a Taylor approximation gives

$$u(x+h,y) = u(x,y) + u_x(x,y)h + u_{xx}(x,y)\frac{h^2}{2} + u_{xxx}(x,y)\frac{h^3}{3!} + u_{xxxx}(\xi_1,y)\frac{h^4}{4!}, \tag{2.2}$$

where ξ_1 lies between x and $x+h$. Substituting h by $-h$ gives

$$u(x-h,y) = u(x,y) - u_x(x,y)h + u_{xx}(x,y)\frac{h^2}{2} - u_{xxx}(x,y)\frac{h^3}{3!} + u_{xxxx}(\xi_2,y)\frac{h^4}{4!}, \tag{2.3}$$

where ξ_2 lies between $x-h$ and x. If we wanted to approximate a first derivative, we would obtain from (2.2) that

$$\frac{u(x+h,y)-u(x,y)}{h} = u_x(x,y) + O(h),$$

and neglecting the error term $O(h)$ gives a first-order finite difference approximation of the first partial derivative of u with respect to x,

$$u_x(x,y) \approx \frac{u(x+h,y)-u(x,y)}{h}.$$

This is called a *forward* approximation. Similarly, we can obtain a *backward* approximation of the first partial derivative with respect to x from (2.3),

$$u_x(x,y) \approx \frac{u(x,y)-u(x-h,y)}{h}.$$

An even better approximation can be obtained using the difference of (2.2) and (2.3), namely,

$$u_x(x,y) = \frac{u(x+h,y)-u(x-h,y)}{2h} + O(h^2),$$

which is a centered approximation that is second-order accurate. One can use the same idea to obtain forward, backward, and centered approximations for the partial derivative with respect to y.

For the second derivative with respect to x, which appears in our Poisson equation, we add (2.2) and (2.3) and obtain

$$u(x+h,y)-2u(x,y)+u(x-h,y) = u_{xx}(x,y)h^2 + (u_{xxxx}(\xi_1,y)+u_{xxxx}(\xi_2,y))\frac{h^4}{4!}.$$

Dividing both sides by h^2, isolating the second derivative term, and assuming that the fourth derivative of u is continuous gives

$$u_{xx}(x,y) = \frac{u(x+h,y) - 2u(x,y) + u(x-h,y)}{h^2} - u_{xxxx}(\xi,y)\frac{h^2}{12}. \tag{2.4}$$

Neglecting the last term on the right-hand side leads to a second-order approximation of the second partial derivative of u with respect to x,

$$u_{xx}(x,y) \approx \frac{u(x+h,y) - 2u(x,y) + u(x-h,y)}{h^2}. \tag{2.5}$$

Similarly, we obtain in the y variable the approximation

$$u_{yy}(x,y) \approx \frac{u(x,y+h) - 2u(x,y) + u(x,y-h)}{h^2}. \tag{2.6}$$

Using these two approximations, we can define a discrete approximation of the Laplace operator.

Definition 2.1. *The* discrete Laplacian Δ_h *is given by*

$$\Delta_h u(x,y) := \frac{u(x+h,y) + u(x,y+h) - 4u(x,y) + u(x-h,y) + u(x,y-h)}{h^2},$$

which is also called the five-point star approximation of the Laplacian.

Applying the discrete Laplacian to u from the Poisson equation (2.1), we get

$$\begin{aligned} \Delta_h u(x,y) &= \frac{u(x+h,y) + u(x,y+h) - 4u(x,y) + u(x-h,y) + u(x,y-h)}{h^2} \\ &= u_{xx}(x,y) + u_{yy}(x,y) + O(h^2) \\ &= f(x,y) + O(h^2). \end{aligned} \tag{2.7}$$

Thus, the solution u of the Poisson equation satisfies the discrete Poisson equation $\Delta_h u(x,y) = f(x,y)$ at each point $(x,y) \in \Omega$ up to a truncation error term $O(h^2)$. The idea of the finite difference method is to neglect the truncation error term and thus to compute an approximation to u at given grid points in the domain. In our case, where the domain Ω is the unit square $(0,1) \times (0,1)$, we discretize the domain with a uniform rectangular mesh with n grid points in each direction. This leads to a mesh size $h = 1/(n+1)$, and the grid is given by $x_i = ih$ and $y_j = jh$ for $i,j = 1,2,\ldots,n$; see Figure 2.2. If n is large, the mesh size h is small, and thus the truncation error $O(h^2)$ should also be small. Neglecting the truncation error in (2.7), denoting by $u_{i,j}$ an approximation of the solution at grid point (x_i, y_j) and letting $f_{i,j} := f(x_i, y_j)$, we obtain the system of equations

$$\Delta_h u_{i,j} = f_{i,j}, \qquad i,j = 1,2,\ldots,n. \tag{2.8}$$

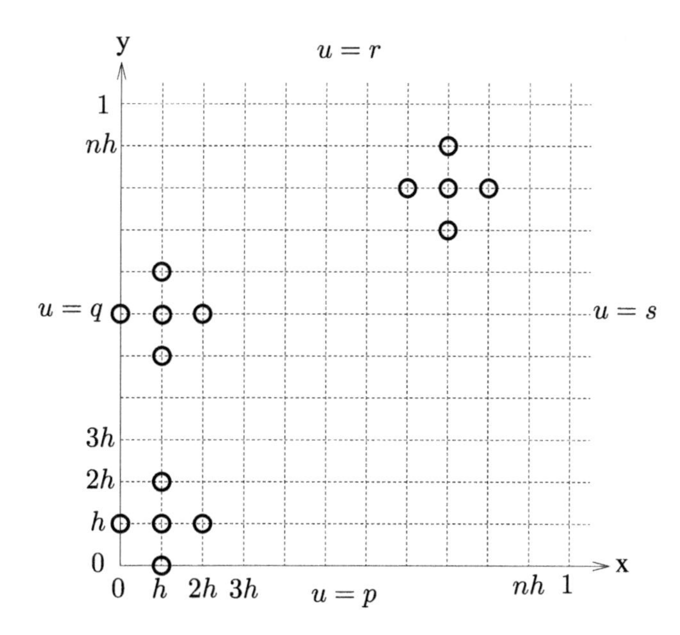

Figure 2.2. *Discretization of the unit square domain* $\Omega = (0,1) \times (0,1)$, *with Dirichlet boundary conditions from* (2.10). *Three five-point stars are shown, two of which involve boundary points.*

Now for the indices $i = 1$, $i = n$, $j = 1$, and $j = n$, the corresponding equations involve the boundary values at $x = 0, 1$ and $y = 0, 1$. More precisely, in the particular case where, for example, $i = 1$ and $j = 1$, as for the five-point star in the lower left corner in Figure 2.2, (2.8) becomes

$$\frac{u_{2,1} + u_{1,2} - 4u_{1,1} + u_{0,1} + u_{1,0}}{h^2} = f_{1,1}, \tag{2.9}$$

where $u_{0,1}$ and $u_{1,0}$ are on the boundary, so their values are given by the boundary condition $u = g$ of problem (2.1). If we denote this boundary condition on each of the sides of the unit square by

$$\left\{ \begin{array}{rcl} u(x,0) & = & p(x), \\ u(0,y) & = & q(y), \\ u(x,1) & = & r(x), \\ u(1,y) & = & s(y), \end{array} \right. \tag{2.10}$$

as shown in Figure 2.2, then (2.9) becomes

$$\frac{u_{2,1} + u_{1,2} - 4u_{1,1} + q_1 + p_1}{h^2} = f_{1,1},$$

where $p_1 := p(x_1)$ and $q_1 := q(y_1)$. Since these boundary values are known, one usually puts them on the right-hand side of the equation and obtains

$$\frac{u_{2,1} + u_{1,2} - 4u_{1,1}}{h^2} = f_{1,1} - \frac{1}{h^2}(q_1 + p_1).$$

Similarly, we obtain for the indices $i = 1$ and $j = 2$ the discrete equation

$$\frac{u_{1,3} + u_{2,2} - 4u_{1,2} + u_{1,1}}{h^2} = f_{1,2} - \frac{1}{h^2} q_2.$$

Continuing this way for all the nodes connected to the boundary, i.e., whenever the index is $i = 1$, $i = n$, $j = 1$, and $j = n$, we gather all equations obtained for every point in the grid in the linear system of equations $A\mathbf{u} = \mathbf{f}$, where the matrix A is given by

$$A = \frac{1}{h^2}\left[\begin{array}{cccc|cccc|cccc|cccc}
-4 & 1 & & & 1 & & & & & & & & & & & \\
1 & -4 & \ddots & & & 1 & & & & & & & & & & \\
 & \ddots & \ddots & 1 & & & \ddots & & & & & & & & & \\
 & & 1 & -4 & & & & 1 & & & & & & & & \\ \hline
1 & & & & -4 & 1 & & & \ddots & & & & & & & \\
 & 1 & & & 1 & -4 & \ddots & & & & & & & & & \\
 & & \ddots & & & \ddots & \ddots & 1 & & & & & & & & \\
 & & & 1 & & & 1 & -4 & & & & \ddots & & & & \\ \hline
 & & & & \ddots & & & & \ddots & & & & 1 & & & \\
 & & & & & & & & & & & & & 1 & & \\
 & & & & & & & & & & & & & & \ddots & \\
 & & & & & & & \ddots & & & & \ddots & & & & 1 \\ \hline
 & & & & & & & & 1 & & & & -4 & 1 & & \\
 & & & & & & & & & 1 & & & 1 & -4 & \ddots & \\
 & & & & & & & & & & \ddots & & & \ddots & \ddots & 1 \\
 & & & & & & & & & & & 1 & & & 1 & -4
\end{array}\right]. \tag{2.11}$$

The matrix A is block tridiagonal and has at most five nonzero entries per row. Such matrices are called *structured*, *sparse* matrices. Although such systems can in principle be solved using Gaussian elimination, the elimination process introduces new nonzero entries, destroying sparsity and incurring high computational and memory costs. An attractive alternative is to use iterative methods, which do not require factorization and are often more efficient for large, sparse problems. We mention in particular Krylov methods [39, 53] which, together with approximate factorizations [53], domain decomposition [47, 60, 54], and multigrid preconditioners [3], lead to the best current iterative solvers for such sparse linear systems; for a simple introduction, see [5].

The vectors $\mathbf{u}$ and $\mathbf{f}$ in the linear system of equations $A\mathbf{u} = \mathbf{f}$ we obtained are

$$\mathbf{u} = \begin{pmatrix} u_{1,1} \\ u_{1,2} \\ \vdots \\ u_{1,n} \\ u_{2,1} \\ u_{2,2} \\ \vdots \\ u_{2,n} \\ \vdots \\ \\ \\ \vdots \\ u_{n,1} \\ u_{n,2} \\ \vdots \\ u_{n,n} \end{pmatrix}, \qquad \mathbf{f} := \begin{pmatrix} f_{1,1} - \frac{1}{h^2}(p_1 + q_1) \\ f_{1,2} - \frac{1}{h^2} q_2 \\ \vdots \\ f_{1,n} - \frac{1}{h^2}(q_n + r_1) \\ f_{2,1} - \frac{1}{h^2} p_2 \\ f_{2,2} \\ \vdots \\ f_{2,n} - \frac{1}{h^2} r_2 \\ \vdots \\ \\ \\ \vdots \\ f_{n,1} - \frac{1}{h^2}(p_n + s_1) \\ f_{n,2} - \frac{1}{h^2}(s_2) \\ \vdots \\ f_{n,n} - \frac{1}{h^2}(r_n + s_n) \end{pmatrix}.$$

Like Runge, we approximated the derivatives in the Poisson equation by finite differences, and thus the finite difference operator is an approximation of the differential operator. However, this does not automatically imply that the solution $\mathbf{u}$ of the large sparse system of equations we obtained is an approximation of the solution u of the Poisson equation (2.1), even though this is what Runge stated without proof.[13] In the next section, we present a convergence analysis of the method, where we will see that the Taylor expansion (2.7) is only one out of three ingredients needed to quantify the error in the discrete solution $\mathbf{u}$.

2.2 ▪ Convergence Analysis

Convergence of the finite difference approximation to the solution of the underlying PDE was first proved by Courant, Friedrich, and Lewy in [9] using a maximum principle and compactness. This was still done in the spirit of proving that the PDE actually has a solution, so nowhere it is assumed that the solution of the PDE exists.[14] The first error estimate was given two years later by Gershgorin [23]; see also the quotes at the beginning of this chapter. Gershgorin proved the result for a general second-order elliptic operator, and we basically follow the same steps in the simpler case of our Poisson equation with Dirichlet boundary conditions on the unit square,

$$\begin{cases} \Delta u = f & \text{in } \Omega = (0,1) \times (0,1), \\ u = g & \text{on } \partial\Omega, \end{cases} \tag{2.12}$$

[13]Runge (1908): "Ist das Netz hinreichend dicht, so werden die gefundenen Werte von u sehr wenig von den wahren Werten verschieden sein, und man kann daraus mit beliebiger Genauigkeit die gesuchte Fläche interpolieren."

[14]Courant, Friedrichs, and Lewy [9]: "Die Lösbarkeit der Differentialgleichungsprobleme setzen wir nirgens voraus; vielmehr erhalten wir durch den Grenzübergang hierfür einen einfachen Beweis."

discretized by a finite difference approximation on the uniform grid given by $x_i = ih$ and $y_j = jh$ for $i, j = 1, 2, \dots, n$ and $h = 1/(n+1)$,

$$\begin{cases} \Delta_h u_{i,j} &= f_{i,j}, \quad i,j = 1,\dots,n, \\ \mathbf{u}_{i,j} &= g_{i,j}, \quad (i,j) \in B, \end{cases} \tag{2.13}$$

where the set of boundary nodes is given by

$$B := \{(i,j) : i = 0, n{+}1 \text{ and } j = 1,\dots,n, \text{ or } j = 0, n{+}1 \text{ and } i = 1,\dots,n\}. \tag{2.14}$$

Following Gershgorin, but as shown for our simpler case in [58], we now prove that the discrete solution $u_{i,j}$ of (2.13) is indeed an approximation of the continuous solution u of (2.12).

Definition 2.2. *For* $\mathbf{u} := \{u_{i,j}\}_{i,j=1}^n$ *defined by* (2.13), *we define the maximum norm in the interior by*

$$||\mathbf{u}||_\infty = \max_{i,j=1,\dots,n} |u_{i,j}|$$

and also a maximum norm on the boundary,

$$||\mathbf{u}||_{\infty,\partial\Omega} = \max\{u_{i,j} : (i,j) \in B\}.$$

In order to prove convergence of the finite difference approximation, we need three main ingredients: a truncation error estimate, a discrete maximum principle, and a Poincaré-type estimate, which bounds the values of a function by its derivatives. These results are given in the following lemmas.

Lemma 2.3 (truncation error estimate). *If the solution u of the Poisson equation* (2.12) *is in* $C^4((0,1)\times(0,1))$ *and satisfies*

$$\left|\frac{\partial^4 u}{\partial x^4}(x,y)\right| \le M_1, \quad \left|\frac{\partial^4 u}{\partial y^4}(x,y)\right| \le M_2 \quad \forall x, y \in (0,1),$$

then $\mathbf{u}$, *defined by* (2.13), *satisfies*

$$||\Delta_h u - \Delta_h \mathbf{u}||_\infty \le \frac{M_1 + M_2}{12} h^2.$$

Proof. Using the definition of the maximum norm, the Taylor expansion (2.4) from the previous subsection, and the fact that $f_{i,j} = f(x_i, y_j)$, we obtain

$$\begin{aligned} ||\Delta_h u - \Delta_h \mathbf{u})||_\infty &= \max_{i,j\in\{1,\dots,n\}} |\Delta_h u(x_i, y_j) - \Delta_h u_{i,j}| \\ &= \max_{i,j\in\{1,\dots,n\}} \left|\Delta u(x_i,y_j) + \tfrac{1}{12}\left(\tfrac{\partial^4 u}{\partial x^4}(\xi_i, y_j) + \tfrac{\partial^4 u}{\partial y^4}(x_i,\eta_j)\right) h^2 - f_{i,j}\right| \\ &= \max_{i,j\in\{1,\dots,n\}} \left|\tfrac{1}{12}\left(\tfrac{\partial^4 u}{\partial x^4}(\xi_i, y_j) + \tfrac{\partial^4 u}{\partial y^4}(x_i,\eta_j)\right) h^2\right| \\ &\le \frac{M_1 + M_2}{12} h^2, \end{aligned}$$

which concludes the proof. $\square$

Lemma 2.4 (discrete maximum principle). *Solutions of the five-point finite difference discretization Δ_h of the Laplace equation satisfy a discrete maximum principle:*

1. *If $\Delta_h \mathbf{u} = 0$, then the approximation $u_{i,j}$ attains its maximum and minimum values on the boundary of the domain, i.e., for $(i,j) \in B$.*

2. *If $\Delta_h \mathbf{u} \leq 0$, then the minimum of $u_{i,j}$ is on the boundary, and if $\Delta_h \mathbf{u} \geq 0$, then the maximum of $u_{i,j}$ is on the boundary.*

Proof. The equation $\Delta_h \mathbf{u} = 0$ implies that for all $i, j = 1, 2, \ldots, n$ of the grid, we have

$$\frac{u_{i+1,j} + u_{i,j+1} - 4u_{i,j} + u_{i-1,j} + u_{i,j-1}}{h^2} = 0.$$

Hence, the numerator must be zero. Solving for $u_{i,j}$, we obtain

$$u_{i,j} = \frac{u_{i+1,j} + u_{i,j+1} + u_{i-1,j} + u_{i,j-1}}{4},$$

which means that $u_{i,j}$ equals the average of its grid neighbors, and so it can be neither larger nor smaller than those neighbors. Thus, no interior value $u_{i,j}$ can be a local maximum or minimum, and it follows that any maxima and minima must be attained on the boundary.

If $\Delta_h \mathbf{u} \leq 0$, following the same reasoning as before, $u_{i,j}$ must be greater than the average of its grid neighbors, and hence it cannot be a local minimum. Similarly, when $\Delta_h \mathbf{u} \geq 0$, $u_{i,j}$ must be smaller than the average of its neighbors, and hence it cannot be a local maximum. □

We call a *grid function* $w_{i,j}$ a discrete function associating a value to every point of the grid.

Lemma 2.5 (Poincaré-type estimate). *For any grid function $w_{i,j}$, $i, j = 0, \ldots, n+1$, such that $w_{i,j} = 0$ on the boundary, i.e., for $(i,j) \in B$, we have*

$$||\mathbf{w}||_\infty \leq \frac{1}{8} ||\Delta_h \mathbf{w}||_\infty.$$

Proof. We consider a particular grid function defined by

$$v_{i,j} = \frac{1}{4}\left(\left(x_i - \frac{1}{2}\right)^2 + \left(y_j - \frac{1}{2}\right)^2\right).$$

Applying the discrete Laplacian to this grid function at any point of the grid $(x_i, y_j) = (ih, jh)$, $i, j = 1, \ldots, n$ and $h = \frac{1}{n+1}$, we obtain

$$\begin{aligned}
\Delta_h v_{i,j} &= \frac{1}{4h^2}\left(\left(ih+h-\frac{1}{2}\right)^2 + \left(ih-h-\frac{1}{2}\right)^2 + 2\left(jh-\frac{1}{2}\right)^2 - 4\left(ih-\frac{1}{2}\right)^2 \ldots\right. \\
&\quad \left. \ldots - 4\left(jh-\frac{1}{2}\right)^2 + \left(jh+h-\frac{1}{2}\right)^2 + \left(jh-h-\frac{1}{2}\right)^2 + 2\left(ih-\frac{1}{2}\right)^2\right) \\
&= \frac{1}{4h^2}\left(2\left(ih-\frac{1}{2}\right)h + h^2 - 2\left(ih-\frac{1}{2}\right)h + h^2 \ldots\right. \\
&\qquad \left. \ldots + 2\left(jh-\frac{1}{2}\right)h + h^2 - 2\left(jh-\frac{1}{2}\right)h + h^2\right) \\
&= 1
\end{aligned} \tag{2.15}$$

independently of i, j. Furthermore, since the grid function $\mathbf{v}$ is a parabola centered at $(\frac{1}{2}, \frac{1}{2})$, it attains its maxima at the corners of the unit square domain, where its value equals $\frac{1}{8}$, and thus we get for the maximum norm on the boundary

$$||\mathbf{v}||_{\infty,\partial\Omega} = \frac{1}{8}. \tag{2.16}$$

Now we consider for any grid function $\mathbf{w}$ the inequality

$$\Delta_h w_{i,j} - ||\Delta_h \mathbf{w}||_\infty \leq 0,$$

which trivially holds since we subtract the maximum over all $i, j = 1, \ldots, n$. Then using (2.15), we multiply the norm term by $1 = \Delta_h v_{i,j}$ and obtain

$$\begin{aligned}
\Delta_h w_{i,j} - ||\Delta_h \mathbf{w}||_\infty &= \Delta_h w_{i,j} - ||\Delta_h \mathbf{w}||_\infty \Delta_h v_{i,j} \\
&= \Delta_h (w_{i,j} - ||\Delta_h \mathbf{w}||_\infty v_{i,j}) \leq 0.
\end{aligned}$$

Now using the discrete maximum principle from Lemma 2.4, we know that the minimum of $w_{i,j} - ||\Delta_h \mathbf{w}||_\infty v_{i,j}$ must occur on the boundary. Since by assumption the grid function $w_{i,j}$ equals zero on the boundary and we found in (2.16) that the maximum value of $\mathbf{v}$ on the boundary is $\frac{1}{8}$, we obtain

$$-||\Delta_h \mathbf{w}||_\infty \frac{1}{8} \leq w_{i,j} - ||\Delta_h \mathbf{w}||_\infty v_{i,j} \leq w_{i,j},$$

where the second inequality holds trivially since $||\Delta_h \mathbf{w}||_\infty v_{i,j} \geq 0$ by definition. With a similar argument for the inequality

$$\Delta_h w_{i,j} + ||\Delta_h \mathbf{w}||_\infty \geq 0,$$

we get the relations

$$||\Delta_h \mathbf{w}||_\infty \frac{1}{8} \geq w_{i,j} + ||\Delta_h \mathbf{w}||_\infty v_{i,j} \geq w_{i,j}.$$

We therefore proved that the grid function $w_{i,j}$ lies in between

$$-\frac{1}{8}||\Delta_h \mathbf{w}||_\infty \leq w_{i,j} \leq \frac{1}{8}||\Delta_h \mathbf{w}||_\infty, \qquad i, j = 1, \ldots, n,$$

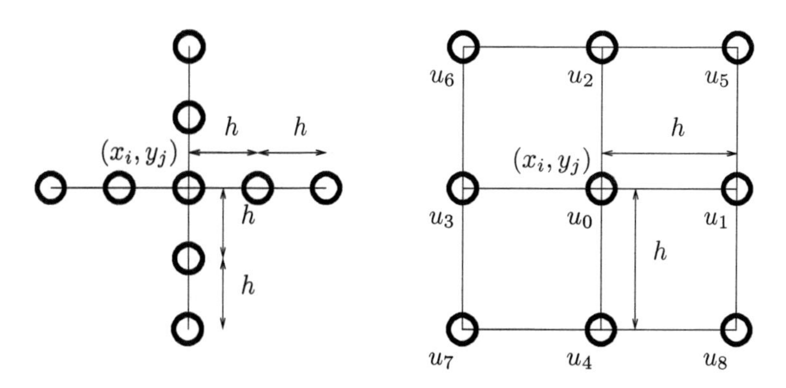

Figure 2.3. *Two nine-point stars for the finite difference approximation of the Laplacian.*

and thus the modulus of the grid function $w_{i,j}$ is bounded by

$$|w_{i,j}| \leq \frac{1}{8}||\Delta_h \mathbf{w}||_\infty, \qquad i, j = 1, \ldots, n.$$

We therefore obtain the norm estimate

$$||\mathbf{w}||_\infty \leq \frac{1}{8}||\Delta_h \mathbf{w}||_\infty$$

as desired. □

Theorem 2.6 (convergence). *Assume that the solution u of the Poisson equation* (2.12) *is in $C^4(\Omega)$. Then the finite difference approximation $\mathbf{u}$ of* (2.13) *converges to u when h tends to zero, and we have the error estimate*

$$||u(x_i, x_j) - u_{i,j}||_\infty \leq Ch^2,$$

where C is a constant and h is the mesh size.

Proof. We simply apply Lemma 2.5 to the norm of the difference and then use Lemma 2.3 to obtain

$$||u(x_i, x_j) - u_{i,j}||_\infty \leq \frac{1}{8}||\Delta_h(u(x_i, x_j) - u_{i,j})||_\infty \leq \frac{M_1 + M_2}{96}h^2,$$

which concludes the convergence proof. □

2.3 ▪ More Accurate Approximations

The five-point star we analyzed in the previous section is, as for all finite difference methods, the result of a Taylor expansion of the unknown solution u at neighboring grid points. This expansion led to the local truncation error estimate in Lemma 2.3. For a more accurate approximation, we would need to cancel more terms in the Taylor expansions, so that the remaining truncation error term is higher order in h. This can be achieved in different ways, two of which are shown in Figure 2.3. For the left choice in

Figure 2.3, the finite difference approximation of the Laplacian is given by

$$\Delta u = \frac{1}{12h^2}(16(u(x+h,y)+u(x,y+h)+u(x-h,y)+u(x,y-h))$$
$$-(u(x+2h,y)+u(x,y+2h)+u(x-2h,y)+u(x,y-2h))-60u(x,y))$$
$$+O(h^4);$$

that is, the local truncation error is of order 4; see Problem 2.3.

To obtain a fourth-order approximation using the right-hand-side scheme in Figure 2.3, we need to include information given by the Poisson equation including the right-hand side. This leads to the finite difference scheme

$$\frac{1}{h^2}(-8(u_1+u_2+u_3+u_4)-2(u_5+u_6+u_7+u_8)+40u_0) = 8f_0+f_1+f_2+f_3+f_4;$$

see Problem 2.3. In this case, we do not get an approximation of the Laplacian independently of the right-hand side but rather a complete discrete system of equations including a particular right-hand side.

2.4 ▪ More General Boundary Conditions

We have already seen how to handle Dirichlet boundary conditions in the finite difference discretization of the Poisson equation; see (2.9). For the one-dimensional Poisson equation $u_{xx} = f$ in $\Omega = (0,1)$ with Dirichlet boundary conditions $u(0) = a$ and $u(1) = b$, this corresponds to solving the linear system

$$\begin{cases} \dfrac{u_{i+1}-2u_i+u_{i-1}}{h^2} &= f(x_i) \quad \forall i = 1,\ldots,n, \\ u_0 &= a, \\ u_{n+1} &= b, \end{cases} \tag{2.17}$$

where $h = \frac{1}{n+1}$. Dirichlet boundary conditions correspond to the prescription of u on the boundary, and in the linear system that results, these boundary values are often transferred to the right-hand side, as we have seen in (2.9).

With *Neumann* boundary conditions, one prescribes the flux through the boundary, which in the case of the Poisson equation is the normal derivative,

$$\frac{\partial u}{\partial \mathbf{n}} = a. \tag{2.18}$$

Hence, the value of u at the boundary is unknown, and an additional equation must be derived to complete the discretized system (2.17). To do so, one discretizes the Neumann boundary condition (2.18) using a finite difference approximation. There are two schemes that can be used:

1. The *one-sided* approach consists of approximating the normal derivative using a one-sided finite difference. For the left boundary, for instance, we would impose

$$\frac{u_1-u_0}{h} = -a,$$

where the minus sign comes from the fact that the outward normal derivative equals minus the derivative with respect to x at the left boundary. Such a discretization is only first-order accurate in general, as one can see using a Taylor expansion,

$$\frac{u(h)-u(0)}{h} = \frac{u(0)+u'(0)h+O(h^2)-u(0)}{h} = u'(0)+O(h).$$

2. A second approach, called the *centered* approach, uses a centered finite difference approximation. On the left boundary, we would have

$$\frac{u_1 - u_{-1}}{2h} = -a. \tag{2.19}$$

This approximation is second-order accurate, just like our discretization of the Laplacian in the interior, since

$$\begin{aligned}\frac{u(h)-u(-h)}{2h} &= \frac{u(0)+u'(0)h+u''(0)h^2/2+O(h^3)-(u(0)-u'(0)h+u''(0)h^2/2-O(h^3))}{2h} \\ &= u'(0) + O(h^2).\end{aligned}$$

We have, however, introduced a *ghost point* $u_{-1} \approx u(-h)$, which lies outside of the domain $\Omega = (0,1)$. In order to obtain an equation for this point, we assume that the discretization of the Laplacian is valid up to the boundary at $x = 0$,

$$\frac{u_1 - 2u_0 + u_{-1}}{h^2} = f_0.$$

Hence, we can solve for the ghost point and obtain

$$u_{-1} = h^2 f_0 + 2u_0 - u_1,$$

which we then introduce into the centered finite difference approximation (2.19) to obtain the new approximation

$$\frac{u_1 - u_0}{h} = -a + \frac{h}{2} f_0.$$

It is interesting to note that if $f_0 = 0$, then the one-sided approach is equivalent to the centered one, and thus both are second-order accurate.

The last commonly used boundary conditions are the *Robin boundary conditions*, in which we prescribe a value to a linear combination of u and its normal derivative. For example, for the left boundary, the condition could be

$$u_x(0) + \alpha u(0) = a.$$

Since we have two choices for approximating the derivative, there are again two discretizations for the Robin boundary conditions:

1. The one-sided finite difference approximation,

$$\frac{u_1 - u_0}{h} + \alpha u_0 = a.$$

2. The centered approximation,

$$\frac{u_1 - u_0}{h} + \alpha u_0 = a + \frac{h}{2} f_0.$$

2.5 ▪ More General Differential Operators

In this section, we consider more general problems of the form

$$\mathcal{L}u = f,$$

where $\mathcal{L}$ is a differential operator, and establish its finite difference approximation, which we denote by $\mathcal{L}_h$.

Definition 2.7. *The finite difference discretization $\mathcal{L}_h$ with mesh parameter h of a differential operator $\mathcal{L}$ of degree m at a point $\mathbf{x} \in \Omega$ is of order p if*

$$\mathcal{L}u(\mathbf{x}) - \mathcal{L}_h u(\mathbf{x}) = O(h^p)$$

for all functions $u \in C^{p+m}(\bar{\Omega})$.

Example 2.8.

1. For the case where $\mathcal{L}$ is just a first derivative in one dimension, $m = 1$,

$$\mathcal{L}u = u_x,$$

we obtain for the one-sided finite difference approximation

$$\mathcal{L}_h u(x) := \frac{u(x+h) - u(x)}{h}.$$

A Taylor expansion reveals that

$$\begin{aligned}\mathcal{L}u - \mathcal{L}_h u &= u_x(x) - \frac{u(x) + u_x(x)h + u_{xx}(\xi)\frac{h^2}{2} - u(x)}{h} \\ &= -u_{xx}(\xi)\frac{h}{2}.\end{aligned}$$

Hence, the order is $p = 1$, and it is necessary to suppose u twice continuously differentiable, i.e., $p + m = 2$.

2. As a second example, we consider the second derivative in one dimension,

$$\mathcal{L}u = u_{xx}.$$

This operator is of degree $m = 2$, and its discrete finite difference approximation we have seen is

$$\mathcal{L}_h u = \frac{u(x+h) - 2u(x) + u(x-h)}{h^2}.$$

Using a Taylor expansion, we have seen that

$$\mathcal{L}u - \mathcal{L}_h u = -\frac{1}{12}u_{xxxx}(\xi)h^2,$$

so that this discretization is of order $p = 2$ and we need the solution u to be four times continuously differentiable, i.e., $p + m = 4$.

We now show several generalizations of finite difference approximations, which go beyond the Poisson equation in two dimensions seen so far. The first generalization is simply to go to three spatial dimensions and consider

$$\begin{cases} \Delta u = u_{xx} + u_{yy} + u_{zz} = f, & \Omega \subset \mathbb{R}^3, \\ u = g, & \partial\Omega. \end{cases}$$

The natural finite difference discretization of this three-dimensional problem is to simply add a finite difference in the third direction, which leads to

$$\Delta_h \mathbf{u} = \frac{u_{i+1,j,k} + u_{i-1,j,k} + u_{i,j+1,k} + u_{i,j-1,k} + u_{i,j,k+1} + u_{i,j,k-1} - 6u_{i,j,k}}{h^2},$$

which is still an order two approximation.

A second generalization is to consider the finite difference discretization of a stationary, nonlinear advection-reaction-diffusion equation, namely,

$$\nu u_{xx}(x) - a u_x(x) + f(u, x) = 0.$$

While the second-order derivative can be discretized as usual by the classical second order finite difference scheme, we have a choice for the first-order derivative: a centered scheme or a one-sided scheme.

- If we choose the centered scheme for the first-order derivative, then we get the discretization

$$\nu \frac{u_{i+1} - 2u_i + u_{i-1}}{h^2} - a\frac{u_{i+1} - u_{i-1}}{2h} + f(u_i, x_i) = 0. \tag{2.20}$$

  This is a second-order scheme for the advection-reaction-diffusion equation.

- For the one-sided scheme, we get the two possible discretizations:

$$\nu \frac{u_{i+1} - 2u_i + u_{i-1}}{h^2} - a\frac{u_i - u_{i-1}}{h} + f(u_i, x_i) = 0, \tag{2.21}$$

$$\nu \frac{u_{i+1} - 2u_i + u_{i-1}}{h^2} - a\frac{u_{i+1} - u_i}{h} + f(u_i, x_i) = 0. \tag{2.22}$$

We show in Figure 2.4 a comparison of the three possible discretizations in (2.20), (2.21), and (2.22) for the linear model problem

$$\nu u_{xx}(x) + u_x(x) = 0, \quad u(0) = 0,\ u(1) = 1,\ \nu = \frac{1}{100}.$$

We make two major observations:

1. When h is not very small, the approximate solution given by the centered finite difference scheme shows unphysical oscillations, which, however, disappear when h becomes smaller. The so-called *upwind scheme* (2.22) always gives a physically correct, monotone solution.

2. The so-called *downwind scheme* (2.21) produces an incorrect solution, even if the mesh size h is refined. The discretization is not convergent. This was already observed in the seminal work by Courant, Friedrichs, and Lewy [9] and led to the so-called CFL condition.[15]

In Figure 2.4, we observe that the downwind scheme is not a convergent scheme, even though the finite difference operator is a first-order discretization. The reason is that the transport term in the advection-diffusion equation, which in the example represents a transport from the right to the left, cannot be taken into account correctly when the finite

[15]Courant, Friedrichs, and Lewy [9]: "...werden wir bei dem Anfangswertproblem hyperbolischer Gleichungen erkennen, dass die Konvergenz allgemein nur dann vorhanden ist, wenn die Verhältnisse der Gittermaschen in verschiedenen Richtungen gewissen Unleichungen genügen, die durch die Lage der Charakteristiken zum Gitter bestimmt werden."

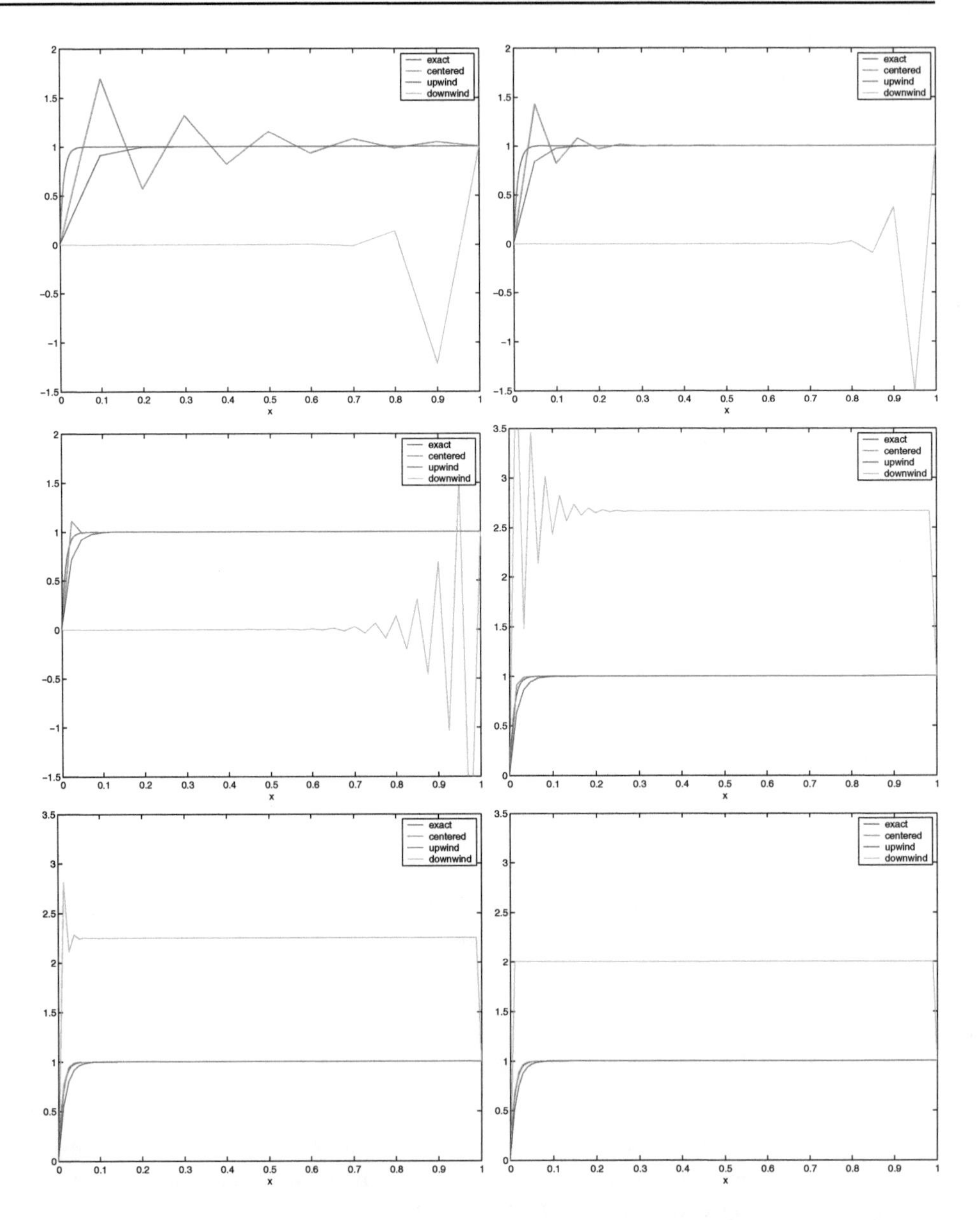

Figure 2.4. *Exact solution and numerical approximation of a simple steady advection-diffusion equation using a centered, an upwind, and a downwind approximation for the advection term for* $h = \frac{1}{10}, \frac{1}{20}, \frac{1}{40}, \frac{1}{60}, \frac{1}{80}, \frac{1}{100}$.

difference discretization looks to the left with a one-sided finite difference. These one-sided discretizations must always be chosen into the direction of where the information is coming from, which is called the upwind direction. If they are chosen correctly, one can show that one obtains the physically correct monotonic solutions, even for coarse mesh sizes; see Problem 2.4. The centered scheme has the advantage that it always looks both ways and hence can capture both flow directions. In addition, it is second-order accurate, which is better than the first-order upwind scheme, as one can see in Figure 2.4. It, however, produces oscillations when the mesh is not fine enough.

We now consider a PDE with variable coefficients,

$$\nu(x)u_{xx} - a(x)u_x + f(u,x) = 0. \tag{2.23}$$

In this case, finite difference schemes can be directly applied; for example, the centered finite difference scheme leads to the discretization

$$\nu(x_i)\frac{u_{i+1} - 2u_i + u_{i-1}}{h^2} - a(x_i)\frac{u_{i+1} - u_{i-1}}{2h} + f(u_i, x_i) = 0,$$

which is again second order. Similarly, one could use an upwind scheme, but then at each grid point, one has to check the sign of $a(x_i)$ and choose the one-sided discretization which looks upwind. Another example with variable coefficients is given by the diffusion equation

$$(a(x)u_x)_x = f,$$

and here it is not immediately clear how to apply a finite difference discretization. One idea is to apply a finite difference scheme twice, once for each first-order derivative. We would therefore first discretize the inner term,

$$a(x_i)u_x(x_i) \approx a(x_i)\frac{u(x_{i+1}) - u(x_i)}{h},$$

and then apply again a one-sided finite difference to the outer derivative in the equation,

$$\begin{aligned}
(a(x_i)u_x(x_i))_x &\approx \frac{1}{h}\left(a(x_i)\frac{u(x_{i+1}) - u(x_i)}{h} - a(x_{i-1})\frac{u(x_i) - u(x_{i-1})}{h}\right) \\
&= \frac{a(x_i)u(x_{i+1}) - (a(x_i) + a(x_{i-1}))u(x_i) + a(x_{i-1})u(x_{i-1})}{h^2}.
\end{aligned}$$

Such an approximation is, however, usually only first-order accurate, as one can check using the Maple commands

```
Lhu:=(a(x)*u(x+h)-(a(x)+a(x-h))*u(x)+a(x-h)*u(x-h))/h^2;
taylor(Lhu,h,4);
```

which gives as a result

$$\begin{aligned}
&(\mathrm{a}(x)\,(\mathrm{D}^{(2)})(u)(x) + \mathrm{D}(a)(x)\,\mathrm{D}(u)(x)) \\
&\quad + \left(-\frac{1}{2}\,\mathrm{D}(a)(x)\,(\mathrm{D}^{(2)})(u)(x) - \frac{1}{2}\,(\mathrm{D}^{(2)})(a)(x)\,\mathrm{D}(u)(x)\right)\,h + \mathrm{O}(h^2),
\end{aligned}$$

and we see that the h-order term in general will not be zero, except for a constant.

Another idea for a finite difference discretization, which is also indicated by the above result in Maple, is to first expand all the derivatives,

$$a_x u_x + a u_{xx} = f,$$

and then to apply again the finite difference discretizations considered in (2.23). This, however, requires the knowledge of the derivative of the diffusion coefficient $a(x)$. We will see in the Chapter 3 on finite volume schemes that there is a further approach to obtain a finite difference scheme for equations with variable coefficients, which is somewhat more natural than the finite difference schemes we have constructed here.

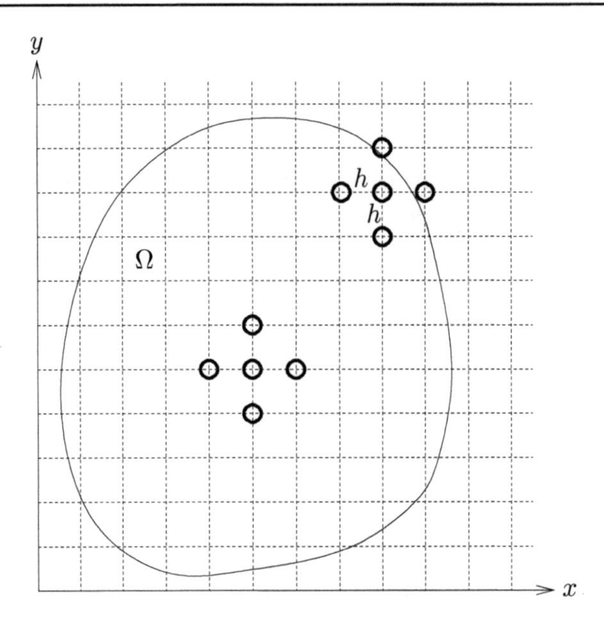

Figure 2.5. *If the domain is not rectangular, the uniform finite difference grid needs to be adapted close to the boundary since grid points will not lie on the boundary in general, as for the top right five-point stencil in this figure.*

2.6 ▪ More General, Nonrectangular Domains

If the domain is not rectangular, such as, for example, in Figure 2.5, there is immediately an additional difficulty in applying a finite difference discretization close to the boundary: A uniform grid will lead to nodes that are not located on the boundary, and thus the standard five-point finite difference discretization of the Laplacian needs to be adapted for the boundary. There are two ways to resolve this problem:

1. We modify the domain slightly, so that its boundary falls precisely on grid points. One can, for example, slightly enlarge or shrink the domain Ω so that the boundary $\partial\Omega_h$ of the new domain Ω_h lies on the grid points of the uniform grid; see Figure 2.6 on the left, where the domain was slightly enlarged. This modification leads in general to a first-order approximation, but it is very easy to implement and is the one implemented in the MATLAB `numgrid` command.

2. A second solution consists of moving the exterior grid points along the regular grid lines until they fall on the boundary $\partial\Omega$ of the domain Ω, as shown in Figure 2.6 on the right. This implies a modification of the finite difference stencil: Since we moved the grid points, we get new local mesh parameters denoted by $\bar{h}$ and $\bar{\bar{h}}$. Using Taylor expansions, we compute now a new finite difference Laplacian for this irregular grid situation. Expanding with the two different mesh parameters h and $\bar{h}$, we find for the x direction

$$u(x+\bar{h}) = u(x) + \bar{h}u'(x) + \frac{\bar{h}^2}{2}u''(x) + \frac{\bar{h}^3}{6}u'''(x) + O(\bar{h}^4),$$
$$u(x-h) = u(x) - hu'(x) + \frac{h^2}{2}u''(x) - \frac{h^3}{6}u'''(x) + O(h^4).$$

Adding the first equation to the second one, which is multiplied with the factor $\frac{\bar{h}}{h}$

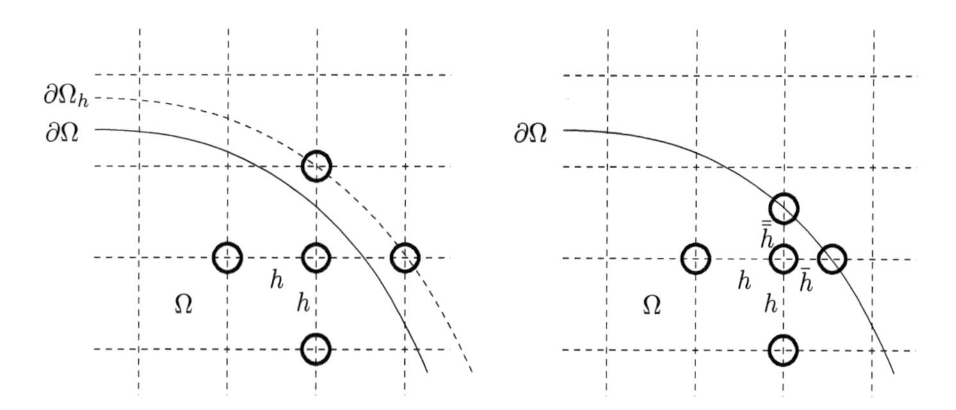

Figure 2.6. *Two options for handling the boundary difficulties caused by a complex geometry of the domain Ω. Left: changing the domain boundary slightly to make it coincide with the grid points. Right: modification of the grid points.*

in order to cancel the first-order derivative terms, we obtain

$$\begin{aligned} &u(x+\bar{h}) + \frac{\bar{h}}{h}u(x-h) \\ &= \left(1+\frac{\bar{h}}{h}\right)u(x) + \left(\frac{\bar{h}^2+h\bar{h}}{2}\right)u''(x) + \left(\frac{\bar{h}^3-\bar{h}h^2}{6}\right)u'''(x) + O(\bar{h}^4) + O(h^4), \end{aligned} \tag{2.24}$$

which leads to the finite difference stencil for the second derivative,

$$\begin{aligned} &\frac{u(x+\bar{h}) - (1+\frac{\bar{h}}{h})u(x) + \frac{\bar{h}}{h}u(x-h)}{\frac{\bar{h}^2+h\bar{h}}{2}} \\ &= u''(x) + \frac{\bar{h}(\bar{h}^2-h^2)}{3\bar{h}(\bar{h}+h)}u'''(x) + O\left(\frac{2\bar{h}^4}{\bar{h}^2+h\bar{h}}\right) + O\left(\frac{2h^4}{\bar{h}^2+h\bar{h}}\right). \end{aligned} \tag{2.25}$$

In order to estimate the error term, it is useful to link the two mesh sizes h and $\bar{h}$ by setting $\bar{h} := \alpha h$, where α is the contraction of the mesh parameter at the boundary. We then obtain for the error term containing $u'''(x)$

$$\frac{\alpha^2h^2 - h^2}{\alpha h + h} = h\frac{\alpha^2-1}{\alpha+1} = O(h) \quad \text{for } \alpha \in (0,1).$$

The same analysis can also be applied in the y direction, and we obtain a modified five-point finite difference stencil for the Laplacian close to the boundary, which is again only first-order accurate.

For nonrectangular domains, the handling of Neumann and Robin boundary conditions is rather difficult when using finite difference discretizations since the normal derivative is no longer in the same direction as the grid lines. We will see that there are more appropriate tools for this case, such as the finite volume method in Chapter 3 and the finite element method in Chapter 5.

2.7 ▪ Room Temperature Simulation Using Finite Differences

We now show how one can use finite differences to solve already quite an interesting problem of steady heat distribution in a room. In doing so, we will resort to many

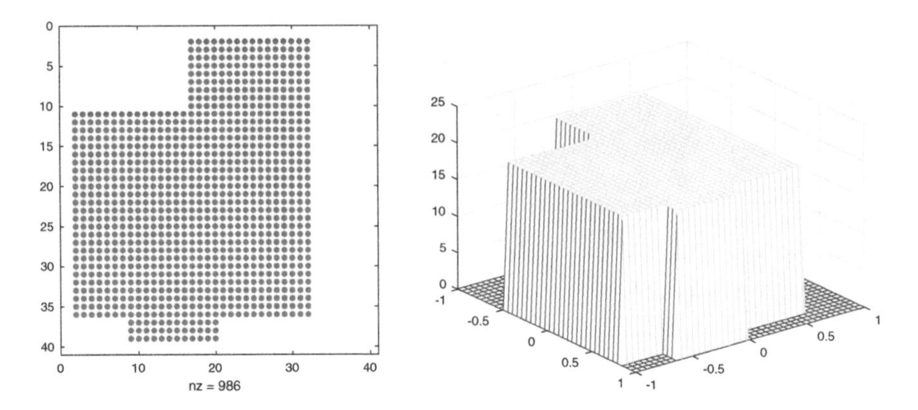

Figure 2.7. *Left: Outline of the living room of Martin Gander's apartment on Durocher Street in Montreal as it appears using the* `spy(G)` *command in MATLAB. Right: Room temperature in summer.*

MATLAB tricks that are employed inside the built-in command `numgrid`. We first show the outline of the living room of Martin Gander on Durocher Street in Montreal, which was depicted already in Figure 1.15 but now as it appears using the `spy(G)` command in MATLAB in Figure 2.7 on the left. The matrix `G` here is part of the temperature simulation program:

```
function U=RoomTemperature(ot,dt,ht,n);
% ROOMTEMPERATURE computes the room temperature in our living room
%   U=RoomTemperature(ot,dt,ht,n); takes the outside temperature ot,
%   the door temperature dt and the heater temperature ht and the
%   number of gridpoints n and then computes the room temperature in
%   our living room

x=linspace(-1,1,n);                               % generate grid
[X,Y]=meshgrid(x,x);
G=((X>-1) & (X<0.6) & (Y>-0.5) & (Y<0.8)) | ...   % area of our living room
  ((X>-0.2) & (X<0.6) & (Y>-1) & (Y<-0.5)) | ...
  ((X>-0.6) & (X<0) & (Y>0.8) & (Y<1));
H=((X>-0.6) & (X<0) & (Y>0.5) & (Y<0.75));        % original heater location
%H=((X>-0.6) & (X<0) & (Y>-0.1) & (Y<0.15));      % heater in the center
%H=((X>0.4) & (X<0.6) & (Y>-0.3) & (Y<0.3));      % heater on the wall
k=find(G);
G=zeros(size(G));                    % Convert from logical to double
G(k)=(1:length(k))';                 % Indices for the matrix
A=delsq(G);                          % Laplacian in the interior
do=[];                               % door indices
for i=2:n-1,                         % add Neumann conditions for insulated
  for j=2:n-1,                       % walls
    no=G(i,j);
    if no~=0,
      if G(i,j-1)==0,                % Neumann condition on the left wall
        A(no,no)=A(no,no)-1;
      end;
      if G(i,j+1)==0,                % Neumann condition on the right wall
        A(no,no)=A(no,no)-1;
      end;
      if G(i+1,j)==0 & i<n-1,        % keep Dirichlet conditions for window
        A(no,no)=A(no,no)-1;
      end;
      if G(i-1,j)==0,
        if (X(i,j)>-0.8 & X(i,j)<-0.3), % keep Dirichlet conditions for door
```

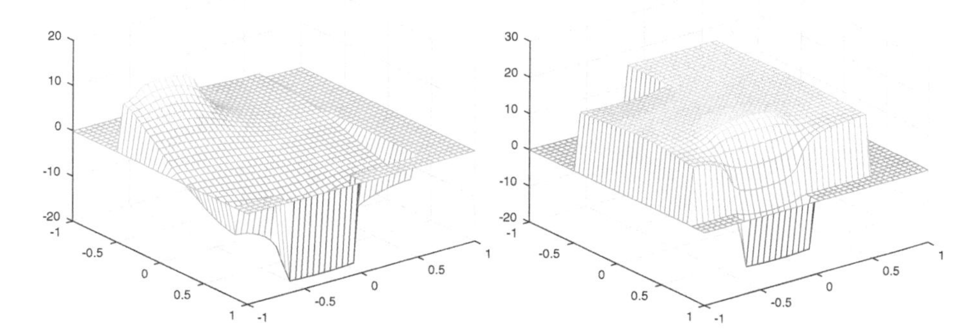

Figure 2.8. *Room temperature in winter without heating (left) and with heating (right), where the heater is well placed next to the window.*

```
          do=[do no];
        else
          A(no,no)=A(no,no)-1;
        end;
      end;
    end;
  end;
end;
h=2/(n-1);
A=-A/h^2;                           % scale the Laplacian with h
b=zeros(length(k),1);
wi=G(end-1,find(G(end-1,:)>0));     % find window indices
he=G(find(H));                      % find heater indices
b(wi)=-1/h^2*ot;                    % add heating and Dirichlet conditions
b(he)=-ht;
b(do)=-1/h^2*dt;
u=A\b;                              % solve using sparse reordered LU
U=G;
U(G>0)=full(u(G(G>0)));             % put solution onto the grid
mesh(X,Y,U);
axis('ij');
```

In the first part of this program, the matrix G is constructed using logical operations to only include regions of a rectangle that are part of the living room. It is best to insert a `keyboard` command to stop the program and to then look at the variables it constructs, i.e., x, y, and G, and to modify the logical operations to create a different living room. Similarly, the position of the heater is also defined in H. Then the discrete Laplace operator is constructed using the `delsq` command in MATLAB. Since this MATLAB command implicitly assumes Dirichlet conditions and most of the walls of the room are insulated, the matrix is modified where needed to contain Neumann conditions. After defining the mesh size h and the right-hand side, the system is solved using backslash and then put on the grid with a compact combination of commands using again the mesh G. With the command

```
RoomTemperature(20,20,0,40);
```

one can obtain the result in Figure 2.7 on the right, with the doors and windows at 20°C and without the heater. Clearly, the temperature is equal everywhere, as one would expect when door and window are held at the same temperature and the rest of the room is fully insulated. On the left in Figure 2.8, we show the distribution in winter

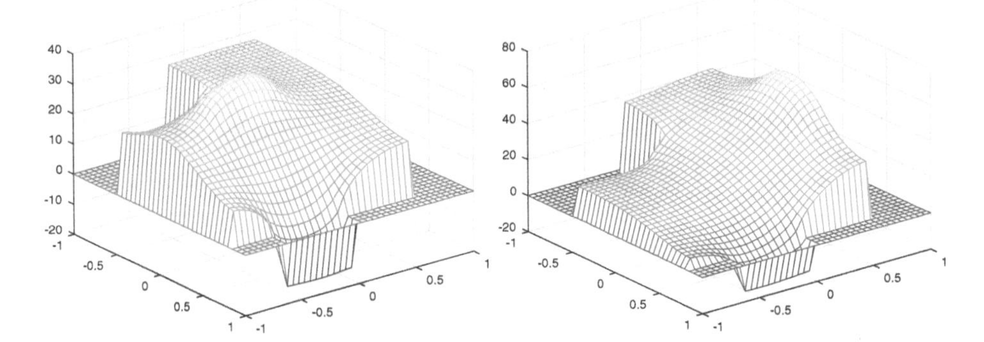

Figure 2.9. *Room temperature in winter with heating, but a poorly placed heater, in the center of the room (left) and on a sidewall (right).*

without the heater, when the door and window are kept at the same temperature as their outside environment, namely at 15 and -20 degrees respectively. The distribution was calculated using the command

```
RoomTemperature(-20,15,0,40);
```

We see that the temperature goes below zero for part of the room. With heating on in winter, we obtain the results in Figure 2.8 on the right, using the command

```
RoomTemperature(-20,15,500,40);
```

The heater is optimally placed close to the window and covering the width of the window. One reaches a comfortable even temperature distribution over the whole room. If the heater had been placed in the center of the room or on a side wall, the temperature would have been significantly worse, as Figure 2.9 shows (just uncomment the corresponding lines in the program `RoomTemperature`).

2.8 ▪ Concluding Remarks

The finite difference method for PDEs is based on an approximation of the differential operators by finite difference operators, which one finds using Taylor expansions. The finite difference method has the following advantages and disadvantages, denoted by plus and minus signs, respectively:

+ The finite difference method is very easy to understand and to program.

– While one can obtain higher-order finite difference stencils, their accuracy will always be a power of h.

– There is no systematic way to derive finite difference approximations, apart from using Taylor expansions and combining terms.

– It is more difficult to discretize problems on general domains, especially if the boundary conditions are not of Dirichlet type.

We will see in the following chapters how each of the negative points can be addressed by different, more sophisticated techniques.

2.9 ▪ Problems

Problem 2.1 (discretization of Neumann boundary conditions).

1. Derive a one-sided finite difference discretization for the treatment of Neumann boundary conditions imposed on a two-dimensional Poisson problem $-\Delta u = f$, where the interior is discretized using the standard five-point stencil. Show that the truncation error is of order one in general.

2. Show how the discretization matrix associated with the five-point finite difference stencil can be easily modified to include Neumann conditions for the case of a rectangular domain.

3. Repeat the above for a centered approximation, and show that the local truncation error is of order two.

4. What is the fundamental difference with the one-dimensional case shown in subsection 2.4?

Problem 2.2 (temperature simulation in a room).

1. Draw a floor plan of your room, including windows, doors, and heaters.

2. Model the temperature in your room using the stationary heat equation $-\Delta u = f$, also called the Poisson equation. To do so, write a MATLAB program similar to the program given in section 2.7. We suppose here that the walls are perfectly insulating, which implies homogeneous Neumann boundary conditions, and for the windows and doors, we suppose no insulation at all, which implies Dirichlet conditions with a given temperature.

3. Compute the room temperature in summer, when the doors and windows are at 20°C. What result do you observe?

4. Compute the room temperature in winter, without heating, on a cold winter day with -10°C outside, i.e., the window is at a temperature of 10°C, and the doors at a temperature of 15°C.

5. Do the same now with the heater turned on such that the temperature is comfortable. Are your heaters well placed?

Problem 2.3 (higher-order finite difference approximations). We have studied the standard five-point finite difference stencil in subsection 2.1. The coefficients of this approximation can be obtained using the Maple commands

```
U[0]:=u(x,y);
U[1]:=u(x+h,y);
U[2]:=u(x,y+h);
U[3]:=u(x-h,y);
U[4]:=u(x,y-h);

delu:=sum(a[i]*U[i],i=0..4);

for i from 0 to 4 do
c[i]:=expand(coeftayl(delu,h=0,i));
od;
```

```
eqs1:=coeffs(c[0],u(x,y));
eqs2:=coeffs(c[1],{D[1](u)(x,y),D[2](u)(x,y)});
eqs3:=coeffs(c[2],{D[1,1](u)(x,y),D[2,2](u)(x,y)});
eqs4:=coeffs(c[3],{D[1,1,1](u)(x,y),D[2,2,2](u)(x,y)});
solve({eqs1,eqs2,eqs3[1]=1,eqs3[2]=1,eqs4},{seq(a[i],i=0..4)});

# gives -4 1 1 1 1 as expected, eqs4 and eqs2 are equivalent
```

Show, using similar commands in Maple, that the nine-point finite difference stencils given in subsection 2.3 have a local truncation error of order four.

Problem 2.4 (discrete maximum principle). The goal is to discover under which conditions there is a discrete maximum principle for the finite difference discretizations (2.20), (2.21), and (2.22) that we have seen in this chapter for the corresponding advection-diffusion equation. We assume for this question that $h > 0$, $a > 0$, and $\nu > 0$.

1. For the centered scheme (2.20), prove that under a condition on the mesh size h, depending on a and ν, the scheme satisfies a discrete maximum principle.
2. Repeat part 1 for (2.21).
3. Repeat part 1 for (2.22).
4. How is the discrete maximum principle related to the monotonicity of the approximate solution?
5. Under the assumptions in this question, which of the two schemes (2.21) and (2.22) is the *upwind* scheme?

Chapter 3

The Finite Volume Method

Perhaps some day in the dim future it will be possible to advance the computations faster than the weather advances and at a cost less than the saving to mankind due to the information gained. But that is a dream.

Lewis F. Richardson, Weather Prediction by Numerical Process, 1922

The high turning angles encountered in high-performance turbines make an orthogonal grid difficult to use and, due to the typical sharp boundary curvatures, increase the danger of computational instability. The foregoing problems have been overcome to a significant extent by the proposed "finite area method" which is a numerical representation of the transient conservation equation in integral form.

P. W. McDonald, The Computation of Transonic Flow through Two-Dimensional Gas Turbine Cascades, 1971

We have seen that the finite difference method has difficulties adapting to nonrectangular geometries with various boundary conditions. This was the main motivation behind the invention of a new technique in the area of computational fluid dynamics; see the second quote above. In 1971, McDonald [43] proposed a new technique now known as the finite volume method. The same technique appeared also independently in 1972 [42], where the main focus was first on splitting methods, but the finite volume method appears toward the end; see Figure 3.1. On a geometry with a mesh as in Figure 3.1 on the left, it is no longer possible to guess difference stencils using Taylor expansions and taking the appropriate combinations. The new idea of the finite volume method is to first integrate the equation over a small so-called *control volume*, then to use the divergence theorem to convert the volume integral into a boundary integral involving fluxes, before finally approximating these fluxes across the boundaries; see Figure 3.1 on the right. This gives a systematic way of constructing finite difference–type stencils on arbitrary meshes with very complicated boundaries and associated boundary conditions, leading to a very flexible method. In the case of a rectangular mesh, one obtains with the finite volume method the same finite difference stencils that we have encountered in Chapter 2 on finite difference methods, and convergence of the method follows as before. For more general meshes, however, the stencils obtained from the finite volume method are not even consistent in the sense of small local truncation errors, as one can simply verify by a Taylor expansion. Finite volume methods are consistent in a different way, namely, that the fluxes are discretized in a conservative manner, and the flux approximations are consistent. We will see that the convergence analysis of finite volume methods requires new technical tools.

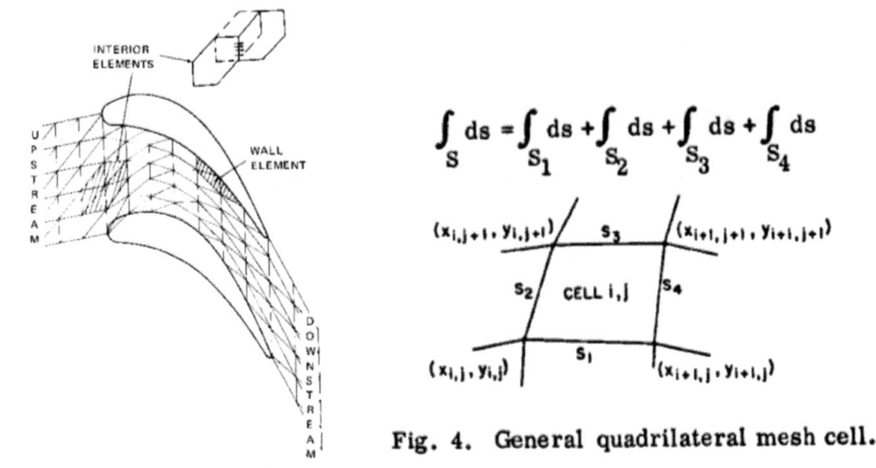

Figure 3.1. *Left: Figure from the original publication of* [43], *indicating why a realistic geometry of a turbine cannot be treated with a rectangular mesh. Reprinted with permission from ASME. Right: Finite volume cell and integration over its four boundaries from the original publication of* [42]. *Reprinted with permission from AIAA.*

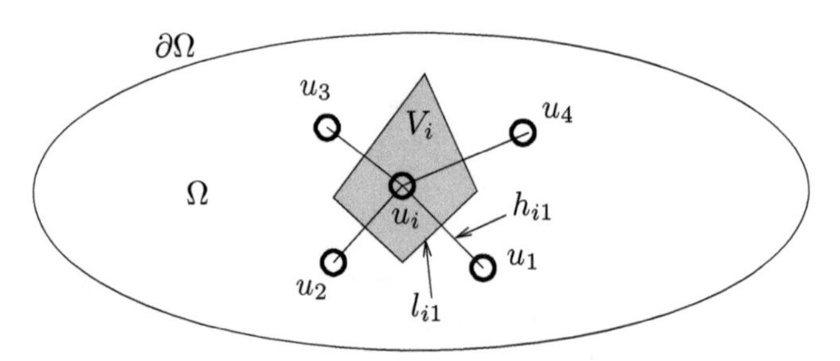

Figure 3.2. *Voronoi cell V_i for the unknown u_i.*

3.1 ▪ Finite Volumes for a General Two-Dimensional Diffusion Equation

We will start by looking at the general diffusion equation

$$\nabla \cdot (a(\mathbf{x})\nabla u) = f \quad \text{in } \Omega \tag{3.1}$$

with a scalar function $a(\mathbf{x})$.

The finite volume method is designed to discretize such differential operators on arbitrary domains with very general grids. Given a grid formed by an arbitrary mixture of polygons with vertices $\{\mathbf{x}_i\} \subset \Omega$, one first has to construct a set of *control volumes* around each $\mathbf{x}_i$, over which the finite volume method then discretizes the PDE. One such construction is to consider Voronoi cells; see Figure 3.2.

Definition 3.1 (Voronoi cell). *For a given set of vertices $\{\mathbf{x}_i\} \subset \Omega$ of a mesh, we define the corresponding* Voronoi cells V_i *by*

$$V_i = \{x \in \Omega : \|\mathbf{x} - \mathbf{x}_i\| \leq \|\mathbf{x} - \mathbf{x}_j\|,\ \forall j \neq i,\ \mathbf{x}_j \textit{ neighbor of } \mathbf{x}_i\}. \tag{3.2}$$

In the finite volume method, one first has to integrate the PDE over each control volume V_i,

$$\int_{V_i} \nabla \cdot (a(\mathbf{x})\nabla u) d\mathbf{x} = \int_{\partial V_i} a(\mathbf{x})\nabla u \cdot \mathbf{n} ds = \int_{V_i} f d\mathbf{x}, \tag{3.3}$$

where we used the divergence theorem to convert the volume integral to a surface integral. Then one approximates the normal derivative $a(\mathbf{x})\nabla u \cdot \mathbf{n} = a(\mathbf{x})\frac{\partial u}{\partial n}$ using the so-called two-point flux approximation (TPFA) scheme: We replace the directional derivative with a finite difference of function values on each side of the control volume, yielding

$$\begin{aligned} \int_{\partial V_i} a(\mathbf{x})\nabla u \cdot \mathbf{n}\, ds &= \sum_{j\sim i} \int_{\Gamma_{ij}} a(\mathbf{x})\nabla u \cdot \mathbf{n}_{ij}\, ds \\ &\approx \sum_{j\sim i} a_{ij} \frac{u_j - u_i}{\|\mathbf{x}_j - \mathbf{x}_i\|} l_{ij} \\ &= \sum_{j\sim i} a_{ij} \frac{u_j - u_i}{h_{ij}} l_{ij} \\ &\overset{!}{=} \text{Vol}(V_i) f_i, \end{aligned} \tag{3.4}$$

where we use the following notation:

h_{ij} is the distance between $\mathbf{x}_i$ and $\mathbf{x}_j$;

a_{ij} is the value of $a(x)$ at the midpoint between $\mathbf{x}_i$ and $\mathbf{x}_j$;

l_{ij} is the length of the boundary part Γ_{ij} of V_i between $\mathbf{x}_i$ and $\mathbf{x}_j$;

$j \sim i$ means j is a neighbor of i.

Collecting the discrete equations

$$\sum_{j\sim i} a_{ij} \frac{u_j - u_i}{h_{ij}} l_{ij} = \text{Vol}(V_i) f_i$$

from (3.4) for all the grid nodes $\mathbf{x}_i$, we obtain a linear system

$$A\mathbf{u} = \mathbf{f}, \tag{3.5}$$

which, as in the case of the finite difference method, is sparse and structured.

Remark 3.1. *Choosing for a_{ij} the value of $a(x)$ at the midpoint between $\mathbf{x}_i$ and $\mathbf{x}_j$ and multiplying with the length l_{ij} is just one possible way to approximate the integral $\int_{\Gamma_{ij}} a(\mathbf{x}) ds$ that remains once the fluxes have been approximated by constants. Any other quadrature formula could also be used.*

3.2 ▪ Boundary Conditions

One of the main advantages of the finite volume method is that it can naturally take boundary conditions into account on very general domains, as we show now.

If the diffusion equation (3.1) is equipped with Dirichlet boundary conditions,

$$u = g \quad \text{on } \partial\Omega, \tag{3.6}$$

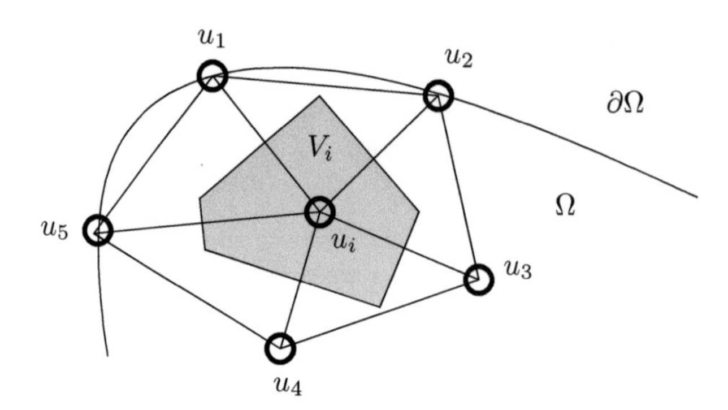

Figure 3.3. *Voronoi cell V_i close to the boundary with a Dirichlet boundary condition.*

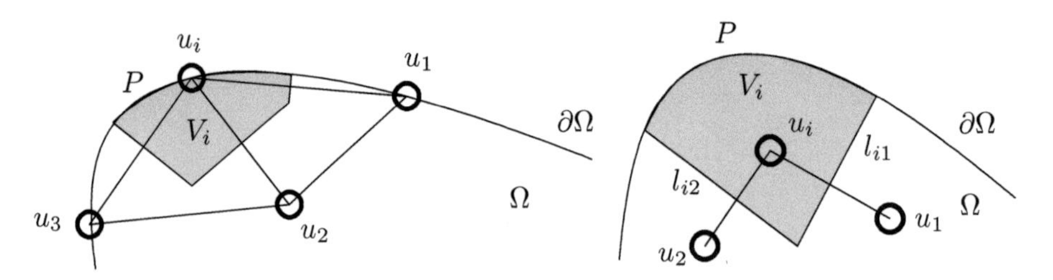

Figure 3.4. *Voronoi cell V_i on the boundary. Left: Vertex-centered case. Right: Cell-centered case.*

it is convenient in the finite volume method to align the mesh such that there are nodes that fall directly onto the boundary, as in Figure 3.3. For a mesh point $\mathbf{x}_i$ next to the boundary, as in Figure 3.3, one then integrates normally over the associated control volume V_i as in (3.4) and obtains for the example in Figure 3.3 the discrete equation

$$a_{i3}\tfrac{u_3-u_i}{h_{i3}}l_{i3}+a_{i4}\tfrac{u_4-u_i}{h_{i4}}l_{i4}+a_{i1}\tfrac{g_1-u_i}{h_{i1}}l_{i1}+a_{i2}\tfrac{g_2-u_i}{h_{i2}}l_{i2}+a_{i5}\tfrac{g_5-u_i}{h_{i5}}l_{i5}=\mathrm{Vol}(V_i)f_i. \tag{3.7}$$

Since for the nodes on the boundary $\partial\Omega$ we know their value from the Dirichlet boundary condition (3.6), $u_j = g_j := g(\mathbf{x}_j)$, $j = 1, 2, 5$, these values can be inserted into the discrete equation (3.7) and then put onto the other side of the equal sign, to be simply included in the right-hand side of the linear system (3.5), very similar to the finite difference case. Note, however, that in the finite volume case, the arbitrary shape of the boundary did not pose any difficulties in the construction of the discrete equation.

We now consider the diffusion equation (3.1) with Neumann boundary conditions,

$$\frac{\partial u}{\partial n} = g \quad \text{on } \partial\Omega. \tag{3.8}$$

To discretize Neumann boundary conditions in the finite volume method, there are two approaches:

vertex centered: the grid points are chosen to lie on the boundary as in the Dirichlet case (see Figure 3.4 on the left);

cell centered: the grid points are all lying inside the domain (see Figure 3.4 on the right).

In the vertex-centered case, where the grid points lie on the boundary, one integrates again over the control volume in the finite volume method and approximates the normal derivatives, except on the real boundary,

$$\begin{aligned}\int_{V_i} \nabla \cdot (a(\mathbf{x})\nabla u) &= \int_{\partial V_i} a(\mathbf{x})\frac{\partial u}{\partial n} ds \\ &\approx a_{i2}\frac{u_2 - u_i}{h_{i2}} l_{i2} + a_{i1}\frac{u_1 - u_i}{h_{i1}} l_{i1} \\ &\quad + a_{i3}\frac{u_3 - u_i}{h_{i3}} l_{i3} + \int_P a(\mathbf{x})\frac{\partial u}{\partial n} ds. \end{aligned} \tag{3.9}$$

In the remaining integral term, the normal derivative is now given by the Neumann boundary condition $\frac{\partial u}{\partial n} = g$ because we integrate on the boundary of the physical domain. This integral can thus be evaluated or approximated by quadrature and the corresponding value put into the right-hand side of the linear system (3.5).

In the cell-centered approach, the control volumes are aligned with the physical boundary of the problem, as shown in Figure 3.4 on the right. The finite volume method can also proceed in this case simply with an integration over the control volume,

$$\begin{aligned}\int_{V_i} \nabla \cdot (a(\mathbf{x})\nabla u) &= \int_{\partial V_i} a(\mathbf{x})\frac{\partial u}{\partial n} ds \\ &\approx a_{i2}\frac{u_2 - u_i}{h_{i2}} l_{i2} + a_{i1}\frac{u_1 - u_i}{h_{i1}} l_{i1} + \int_P a(\mathbf{x})\frac{\partial u}{\partial n} ds.\end{aligned}$$

Since the remaining integral is entirely along the Neumann boundary, it can be evaluated from the prescribed fluxes and incorporated into the right-hand side of the discrete linear system.

To discretize a *Robin* boundary condition of the type

$$a(\mathbf{x})\frac{\partial u}{\partial n} + \gamma u = g,$$

we can again consider a vertex-centered and a cell-centered approach. The vertex-centered approach is similar to the Neumann case, except we substitute the Robin condition into the last integral in (3.9),

$$\begin{aligned}\int_{V_i} \nabla \cdot (a(\mathbf{x})\nabla u) &\approx \sum_{j=1}^{3} a_{ij}\frac{u_j - u_i}{h_{ij}} l_{ij} + \int_P a(\mathbf{x})\frac{\partial u}{\partial n} ds \\ &= \sum_{j=1}^{3} a_{ij}\frac{u_j - u_i}{h_{ij}} l_{ij} + \int_P (g - \gamma u(\mathbf{x})) ds \\ &\approx \sum_{j=1}^{3} a_{ij}\frac{u_j - u_i}{h_{ij}} l_{ij} - \gamma u_i \cdot l_P + \int_P g ds,\end{aligned} \tag{3.10}$$

where l_P is the length of P. In the cell-centered case, however, we have the further complication that $\int_P \gamma u(\mathbf{x}) ds$ cannot be directly approximated in terms of u_i since the grid point does not lie on the boundary. To approximate this integral, we introduce a *ghost point* with value u_g outside the domain such that the segment between u_g and u_i is normal to P and its midpoint lies on P; see Figure 3.5. Then we can introduce the approximation

$$\int_P \gamma u(\mathbf{x}) ds \approx \gamma \left(\frac{u_i + u_g}{2}\right) l_P.$$

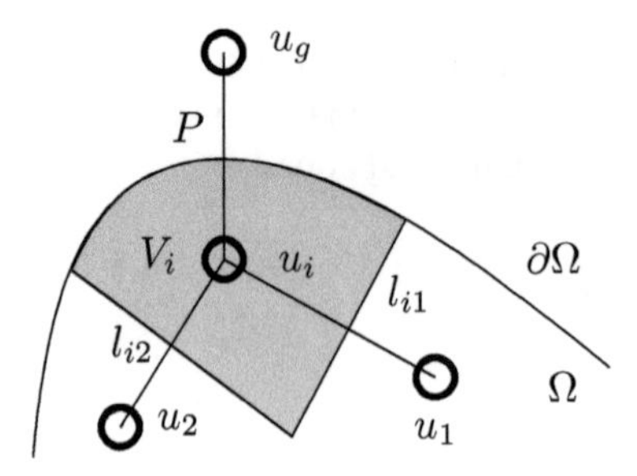

Figure 3.5. *Voronoi cell V_i on the boundary for a Robin boundary condition in the cell-centered case.*

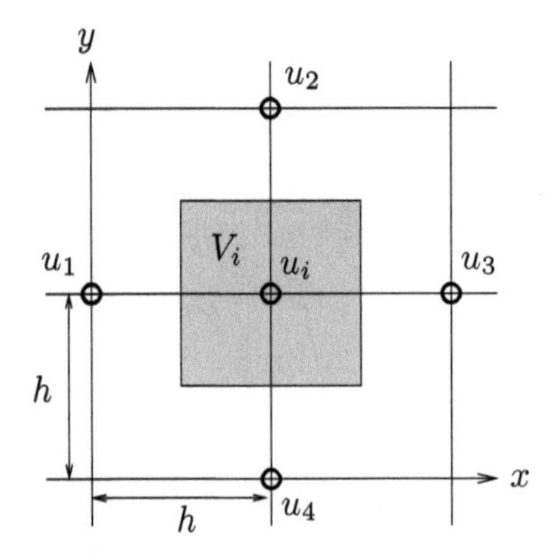

Figure 3.6. *Cartesian square mesh to investigate the relation between the finite volume and the finite difference method.*

Just like for finite difference methods, the ghost point unknown u_g can be eliminated using the boundary condition

$$g = a\frac{\partial u}{\partial n} + \gamma u \approx a\frac{u_g - u_i}{h_{gi}} + \gamma\frac{u_i + u_g}{2},$$

where g and a are to be evaluated at the midpoint of u_g and u_i.

3.3 ▪ Relation between Finite Volumes and Finite Differences

We have seen that the finite volume method also constructs discretization stencils, such as the classical five-point star we had investigated for the finite difference method, but on arbitrary meshes. It is thus of interest to see what kind of discretization stencil the finite volume method produces on a regular grid. To investigate this, we consider the Poisson equation

$$\Delta u = f \quad \text{in } \Omega = (0,1)^2 \tag{3.11}$$

and choose for our mesh a simple square Cartesian mesh, as indicated in Figure 3.6.

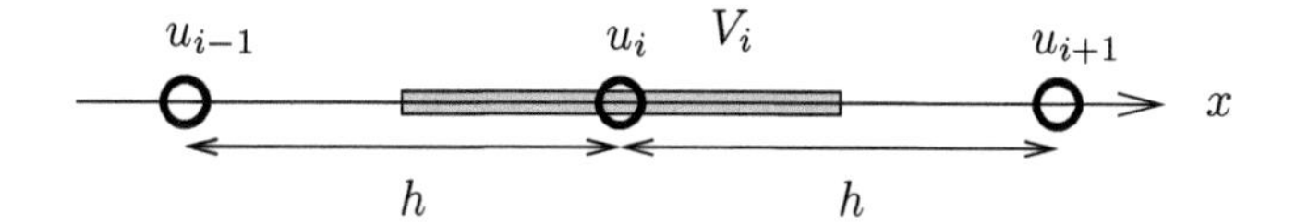

Figure 3.7. *Control volume V_i in one dimension.*

Integrating over the control volume as usual in the finite volume method, we find

$$\int_{V_i} u d\mathbf{x} = \int_{\partial V_i} \frac{\partial u}{\partial n} ds$$
$$\approx \sum_{j=1}^{4} \frac{u_j - u_i}{h} h \overset{!}{=} h^2 f_i, \tag{3.12}$$

which gives for the finite volume method the difference stencil

$$\Rightarrow u_1 + u_2 + u_3 + u_4 - 4u_i = h^2 f_i. \tag{3.13}$$

This is equivalent to the classical five-point finite difference stencil we have seen earlier, as one only needs to divide by h^2. As a result, we see that the finite volume method is also second-order accurate in this case.

Because of this intimate relationship between finite differences and finite volumes, it is often convenient, when working with finite differences, to construct stencils using finite volume techniques, especially near irregular boundaries or in the presence of variable coefficients.

Example 3.2. We consider again the example that we tried to discretize in Chapter 2 on finite differences,

$$(a(x)u_x)_x = f.$$

With the finite volume method, using a control volume as indicated in Figure 3.7, we obtain

$$\int_{V_i} (au_x)_x dx = a(x)u_x(x)\Big|_{x_i - h/2}^{x_i + h/2} \approx a\left(x_i + \frac{h}{2}\right)\frac{u_{i+1} - u_i}{h} - a\left(x_i - \frac{h}{2}\right)\frac{u_i - u_{i-1}}{h}$$
$$= \frac{a(x_i + \frac{h}{2})u_{i+1} - (a(x_i + \frac{h}{2}) + a(x_i - \frac{h}{2}))u_i + a(x_i - \frac{h}{2})u_{i-1}}{h}$$
$$\overset{!}{=} h f_i.$$

We therefore obtain the finite difference stencil

$$\frac{a(x_i + \frac{h}{2})u_{i+1} - (a(x_i + \frac{h}{2}) + a(x_i - \frac{h}{2}))u_i + a(x_i - \frac{h}{2})u_{i-1}}{h^2} = f_i. \tag{3.14}$$

It would have been difficult to guess this just from writing Taylor expansions as in finite differences. To check the truncation error of the new stencil (3.14), we can, however, use Taylor expansions. Expanding the terms appearing in the new stencil, we get

$$\begin{aligned}
a\left(x_i + \frac{h}{2}\right) &= a(x_i) + a_x\frac{h}{2} + a_{xx}\frac{h^2}{8} + a_{xxx}\frac{h^3}{8\cdot 3!} + \mathcal{O}(h^4),\\
a\left(x_i - \frac{h}{2}\right) &= a(x_i) - a_x\frac{h}{2} + a_{xx}\frac{h^2}{8} - a_{xxx}\frac{h^3}{8\cdot 3!} + \mathcal{O}(h^4),\\
u(x_i+h) &= u(x_i) + u_x h + u_{xx}\frac{h^2}{2} + u_{xxx}\frac{h^3}{3!} + \mathcal{O}(h^4),\\
u(x_i-h) &= u(x_i) - u_x h + u_{xx}\frac{h^2}{2} - u_{xxx}\frac{h^3}{3!} + \mathcal{O}(h^4).
\end{aligned}$$

This implies that

$$-\left(a\left(x_i + \frac{h}{2}\right) + a\left(x_i - \frac{h}{2}\right)\right) = -\left(2a(x_i) + a_{xx}(x_i)\frac{h^2}{4} + \mathcal{O}(h^4)\right)$$

and also that

$$\begin{aligned}
a\left(x_i+\frac{h}{2}\right)u(x_i+h)+a\left(x_i-\frac{h}{2}\right)u(x_i-h) &= a(x_i)\left(2u(x_i)+u_{xx}(x_i)h^2+\mathcal{O}(h^4)\right)\\
&\quad + a_x\frac{h}{2}(2u_x(x_i)h) + \mathcal{O}(h^4)\\
&\quad + a_{xx}\frac{h^2}{8}(2u(x_i)) + \mathcal{O}(h^4).
\end{aligned}$$

Introducing these results into the discrete equation (3.14), we obtain

$$\begin{aligned}
\frac{1}{h^2}(a_x u_x h^2 + a u_{xx} h^2 + \mathcal{O}(h^4)) &= a_x u_x + a u_{xx} + \mathcal{O}(h^2)\\
&= (a(x)u_x(x))_x + \mathcal{O}(h^2),
\end{aligned}$$

and we see that the finite volume method automatically led to a truncation error of order two.

The relation between the finite volume and the finite difference method can be used to analyze the convergence of the finite volume method when the stencil is equivalent to a finite difference stencil whose convergence properties are known. However, we will see in the next sections that finite volume methods also converge in much more general settings; in particular, convergence holds even when the method is inconsistent in the sense of Lemma 2.3, i.e., when the truncation error does not shrink to zero as the mesh is refined.

3.4 ▪ Finite Volume Methods Are Not Consistent

To show the main ideas, we consider first a one-dimensional model problem,

$$-u_{xx} = f \quad \text{in } (0,1) \quad u(0) = u(1) = 0. \tag{3.15}$$

We discretize the domain with a very general mesh with primal mesh points x_i and dual mesh points $x_{i+\frac{1}{2}}$,

$$0 = x_0 = x_{\frac{1}{2}} < x_1 < x_{\frac{3}{2}} < \cdots < x_{N+\frac{1}{2}} = x_{N+1} = 1; \tag{3.16}$$

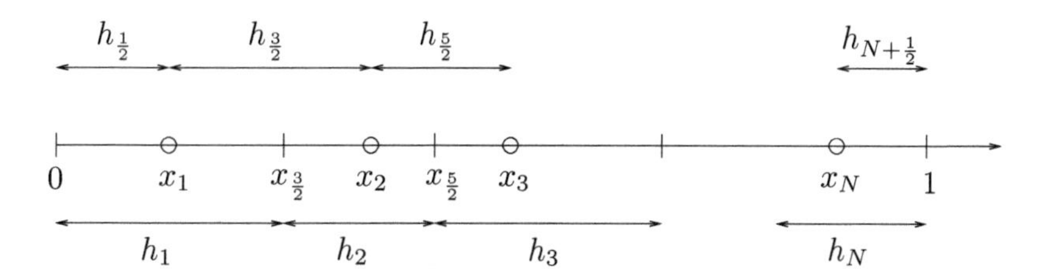

Figure 3.8. *General finite volume mesh with primal nodes x_i and dual nodes $x_{i+\frac{1}{2}}$. Note that we set $x_0 = x_{\frac{1}{2}} = 0$ and $x_{N+\frac{1}{2}} = x_{N+1} = 1$ at the end points.*

see Figure 3.8, where we also indicate the primal mesh sizes $h_i := x_{i+\frac{1}{2}} - x_{i-\frac{1}{2}}$ and the dual mesh sizes $h_{i+\frac{1}{2}} := x_{i+1} - x_i$. The finite volume cell associated with the primal mesh point x_i is $(x_{i-\frac{1}{2}}, x_{i+\frac{1}{2}})$, and integrating (3.15) on such a cell, we get

$$\int_{x_{i-\frac{1}{2}}}^{x_{i+\frac{1}{2}}} -u_{xx}(x)dx = -\left(u_x(x_{i+\frac{1}{2}}) - u_x(x_{i-\frac{1}{2}})\right) = \int_{x_{i-\frac{1}{2}}}^{x_{i+\frac{1}{2}}} f(x)dx. \tag{3.17}$$

If we now approximate the fluxes $u_x(x_{i\pm\frac{1}{2}})$ by finite differences as usual in the finite volume method, we obtain the scheme

$$-\left(\frac{u_{i+1}-u_i}{h_{i+\frac{1}{2}}} - \frac{u_i - u_{i-1}}{h_{i-\frac{1}{2}}}\right) = h_i f_i, \quad i = 1, 2, \dots, N, \tag{3.18}$$

where $u_0 = u_{N+1} = 0$, and we defined

$$f_i := \frac{1}{h_i}\int_{x_{i-\frac{1}{2}}}^{x_{i+\frac{1}{2}}} f(x)dx.$$

Following what we have learned for finite difference methods, we first investigate the truncation error of this scheme. To do so, we insert the exact solution into the scheme (3.18) after having divided by h_i to obtain a finite difference type equation with f_i on the right-hand side and compute the residual,

$$\begin{aligned}
r_i &:= -\tfrac{1}{h_i}\left(\tfrac{u(x_{i+1})-u(x_i)}{h_{i+\frac{1}{2}}} - \tfrac{u(x_i)-u(x_{i-1})}{h_{i-\frac{1}{2}}}\right) - f_i \\
&= -\tfrac{1}{h_i}\left(\tfrac{u_x(x_i)h_{i+\frac{1}{2}} + \frac{1}{2}u_{xx}(x_i)h^2_{i+\frac{1}{2}} + O(h^3_{i+\frac{1}{2}})}{h_{i+\frac{1}{2}}} - \tfrac{u_x(x_i)h_{i-\frac{1}{2}} - u_{xx}(x_i)h^2_{i-\frac{1}{2}} + O(h^3_{i-\frac{1}{2}})}{h_{i-\frac{1}{2}}}\right) - f_i \\
&= -\tfrac{1}{h_i}\left(\tfrac{h_{i+\frac{1}{2}}}{2}u_{xx}(x_i) + O(h^2_{i+\frac{1}{2}}) + \tfrac{h_{i-\frac{1}{2}}}{2}u_{xx}(x_i) + O(h^2_{i-\frac{1}{2}})\right) - f_i \\
&= -\tfrac{h_{i+\frac{1}{2}}+h_{i-\frac{1}{2}}}{2h_i}u_{xx}(x_i) - f_i + \left(O(h^2_{i+\frac{1}{2}}) + O(h^2_{i-\frac{1}{2}})\right).
\end{aligned}$$

To see under which conditions the local truncation error can be made small, we use again the trick of setting $h_{i+\frac{1}{2}} = \alpha h_i$ and $h_{i-\frac{1}{2}} = \beta h_i$, which leads to

$$r_i = -\frac{\alpha+\beta}{2}u_{xx}(x_i) - f(x_i) + O(h_i), \tag{3.19}$$

where we used that $f_i = f(x_i) + O(h_i)$; see the definition (3.18). From (3.19), we see that the residual becomes $O(h_i)$ if and only if $\alpha + \beta = 2$, when the $O(1)$ term vanishes because of the PDE $-u_{xx}(x_i) = f(x_i)$.

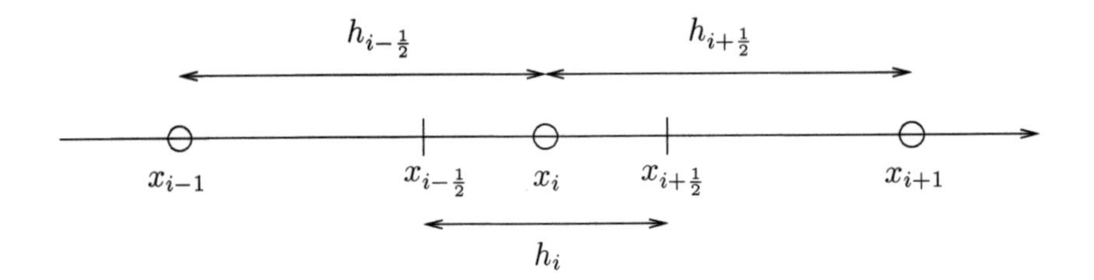

Figure 3.9. *Example of an inconsistent finite volume mesh.*

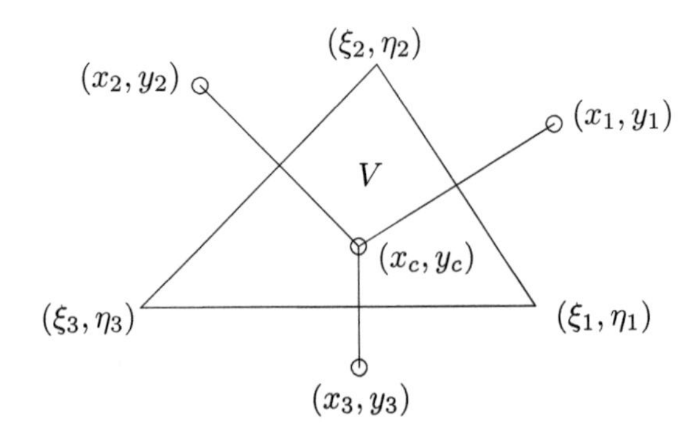

Figure 3.10. *Example of an inconsistent finite volume mesh.*

Example 3.3. Suppose we have a mesh satisfying $h_{i+\frac{1}{2}} = h_{i-\frac{1}{2}}$, but the cell centered at x_i is chosen such that $h_i = \frac{2}{3}h_{i+\frac{1}{2}}$; see Figure 3.9. Then we have $h_{i\pm\frac{1}{2}} = \frac{3}{2}h_i$, which implies that $\alpha = \beta = \frac{3}{2}$. Thus, we have $\alpha + \beta = 3 \neq 2$, so the finite volume scheme is not consistent.

One might argue that in this one-dimensional example, it does not make sense to place the finite volume cell boundaries $x_{i+\frac{1}{2}}$ anywhere other than midway between the x_i. However, inconsistent finite volume schemes arise much more naturally in two dimensions, even when we put the boundaries halfway between primal nodes to obtain a Voronoi cell: Such inconsistencies are in fact the typical case, as we now show for the classical *FV4 finite volume scheme*.

FV4 stands for finite volume scheme with four nodes; i.e., the Voronoi cell is a triangle, as shown in Figure 3.10. This is also often referred to as the two-point flux approximation (TPFA) scheme. To compute the FV4 stencil for the model equation

$$-\Delta u = f,$$

we integrate the equation over the Voronoi cell V and obtain

$$-\int_V u_{xx} dx = \int_{\partial V} \frac{\partial u}{\partial n} = \int_V f dx.$$

Approximating the fluxes by finite differences, we get the FV4 scheme,

$$\sum_{i=1}^{3} \frac{u_i - u_c}{\sqrt{(x_i - x_c)^2 + (y_i - y_c)^2}} \sqrt{(\xi_{i+1} - \xi_i)^2 + (\eta_{i+1} - \eta_i)^2} = \text{vol}(V) f_i, \quad (3.20)$$

where we let $f_i := \frac{1}{\text{vol}(V)} \int_v f(x)dx$ and define for convenience $\xi_4 := \xi_1$ and $\eta_4 := \eta_1$ so we do not have any special cases. To check whether this scheme is consistent, we need to compute for a given set of points (x_i, y_i) ordered counterclockwise around the center point (x_c, y_c) the corresponding corners of the Voronoi cell (ξ_i, η_i), which satisfy the equation

$$(x_i-\xi_i)^2+(y_i-\eta_i)^2 = (x_{i-1}-\xi_i)^2+(y_{i-1}-\eta_i)^2 = (x_c-\xi_i)^2+(y_c-\eta_i)^2, \quad i = 1, 2, 3,$$

where we also defined for convenience $x_0 := x_3$ and $y_0 := y_3$. Inserting the solutions into the FV4 scheme and using a Taylor expansion to check for consistency is best done using Maple. The following Maple commands compute, for an arbitrary number of neighboring nodes N ordered counterclockwise around a center node (x_c, y_c), the corresponding Voronoi cell and its associated finite volume stencil:

```
N:=4;
x[0]:=x[N]; y[0]:=y[N];
for i from 1 to N do
  eq:={(x[i]-xi[i])^2+(y[i]-eta[i])^2=(x[i-1]-xi[i])^2+(y[i-1]-eta[i])^2,
    (x[i]-xi[i])^2+(y[i]-eta[i])^2=(xc-xi[i])^2+(yc-eta[i])^2};
  solve(eq,{xi[i],eta[i]});
  assign(%);
od;
xi[N+1]:=xi[1]; eta[N+1]:=eta[1];
i:='i';s:=sum((u(x[i],y[i])-u(xc,yc))/sqrt((x[i]-xc)^2+(y[i]-yc)^2)
  *sqrt((xi[i+1]-xi[i])^2+(eta[i+1]-eta[i])^2),i=1..N);
```

Next, we introduce the generic mesh size h in order to easily expand in a Taylor series and also set the center node to $(0, 0)$ for simplicity:

```
for i from 1 to N do
  x[i]:=a[i]*h; y[i]:=b[i]*h;
od;
xc:=0; yc:=0;
```

We first choose the parameters to obtain the standard five-point finite difference stencil, plot it, and compute the truncation error:

```
a[1]:=1;b[1]:=0;a[2]:=0;b[2]:=1;a[3]:=-1;b[3]:=0;a[4]:=0;b[4]:=-1;
i:='i';h:=1/10;P1:=plot([seq(xi[i],i=1..N+1)],[seq(eta[i],i=1..N+1)]);
P2:=plot([seq(x[i],i=1..N)],[seq(y[i],i=1..N)],style=point,
  scaling=constrained);
plots[display](P1,P2);
h:='h';simplify(series(s,h,4));
```

We obtain as expected the consistent discretization of the Laplacian that has a truncation error of order two (note that a factor h^2 is also present on the right-hand side from the integration over the finite volume cell of size $h \times h$),

$$((D_{1,1}(u)(0,0) + (D_{2,2}(u)(0,0))\, h^2 + O(h^4),$$

and we show the resulting stencil plot from Maple in Figure 3.11 on the left. However, if we change the nodes a little using the Maple commands

```
a[1]:=1;b[1]:=1/10;a[2]:=-1/10;b[2]:=1;a[3]:=-1;b[3]:=1/5;
a[4]:=-1/5;b[4]:=-1;
i:='i';h:=1/10;P1:=plot([seq(xi[i],i=1..N+1)],[seq(eta[i],i=1..N+1)]);
P2:=plot([seq(x[i],i=1..N)],[seq(y[i],i=1..N)],style=point,
  scaling=constrained);
plots[display](P1,P2);
h:='h';simplify(series(s,h,3));
```

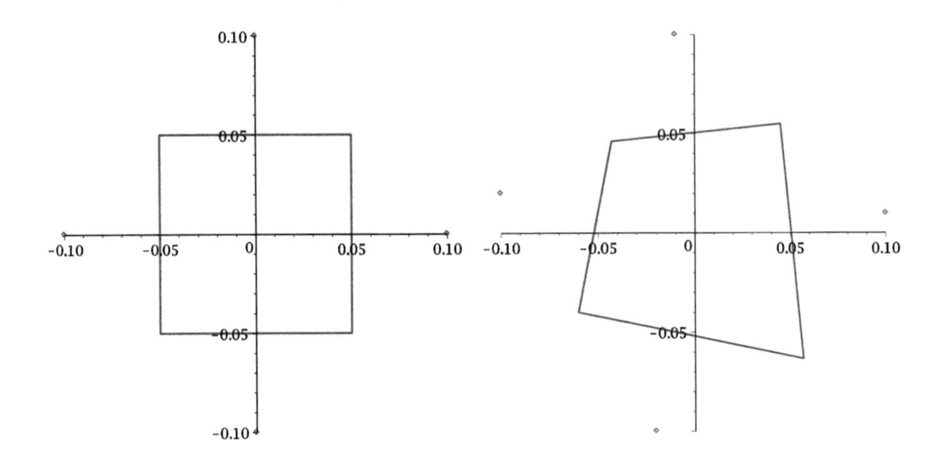

Figure 3.11. *Left: Standard five-point finite difference stencil obtained with a finite volume method, which is consistent. Right: Irregular five-point finite difference stencil obtained with the finite volume method, which leads to an inconsistent discretization.*

we obtain an inconsistent discretization of the Laplacian,

$$\left(\frac{1471}{1400}D_{1,1}(u)(0,0) + \frac{9}{98}D_{1,2}(u)(0,0) + \frac{2563}{2450}D_{2,2}(u)(0,0)\right) h^2 + O(h^3)$$

with the associated stencil plot from Maple in Figure 3.11 on the right. From what we have seen in Chapter 2 on finite differences, this method would not seem to have any chance of being convergent since a small truncation error was the essential ingredient to prove convergence of the finite difference method. A new approach is therefore needed to understand the convergence of finite volume methods.

3.5 ▪ Convergence Analysis

The general ideas of this new type of convergence analysis can be understood in the simple one-dimensional example (3.15) discretized on the very general mesh (3.16), which led to the finite volume scheme (3.18). For a more general treatment, see [17], on which the following one-dimensional result is based.

Theorem 3.4 (FV error estimate). *Assume that in the problem* (3.15) *the right-hand side f is in $C^1([0,1])$, and that the solution u is in $C^2([0,1])$. Then there exists a solution $\mathbf{u} := (u_1, u_2, \ldots, u_N)$ of the finite volume discretization* (3.18) *on the mesh* (3.16), *and the error $e_i := u(x_i) - u_i$ satisfies the error estimates*

$$\sum_{i=0}^{N} \frac{(e_{i+1} - e_i)^2}{h_{i+\frac{1}{2}}} \leq C^2 h^2 \tag{3.21}$$

and

$$|e_i| \leq Ch, \quad i = 1, 2, \ldots, N, \tag{3.22}$$

where C represents some constant that depends only on the solution u and $h := \max_i h_i$, $i = \frac{1}{2}, 1, \frac{3}{2}, \ldots, N, N + \frac{1}{2}$.

Proof. We first multiply (3.18) by u_i and obtain

$$\frac{u_i^2 - u_i u_{i-1}}{h_{i-\frac{1}{2}}} - \frac{u_{i+1}u_i - u_i^2}{h_{i+\frac{1}{2}}} = h_i u_i f_i.$$

We now sum this equation from $i = 1$ to N. Using the fact that $u_0 = u_{N+1} = 0$ and regrouping the terms, we obtain

$$\begin{aligned}
&\frac{u_1^2}{h_{\frac{1}{2}}} - \frac{u_2u_1 - u_1^2}{h_{\frac{3}{2}}} + \frac{u_2^2 - u_2u_1}{h_{\frac{3}{2}}} - \frac{u_3u_2 - u_2^2}{h_{\frac{5}{2}}} + \frac{u_3^2 - u_3u_2}{h_{\frac{5}{2}}} - \cdots + \frac{u_N^2}{h_{N+\frac{1}{2}}}\\
&= \frac{u_1^2}{h_{\frac{1}{2}}} + \frac{(u_2-u_1)^2}{h_{\frac{3}{2}}} + \frac{(u_3-u_2)^2}{h_{\frac{3}{2}}} + \cdots + \frac{u_N^2}{h_{N+\frac{1}{2}}},
\end{aligned}$$

which leads to the closed-form summation formula

$$\frac{u_1^2}{h_{\frac{1}{2}}} + \sum_{i=1}^{N-1} \frac{(u_{i+1}-u_i)^2}{h_{i+\frac{1}{2}}} + \frac{u_N^2}{h_{N+\frac{1}{2}}} = \sum_{i=1}^{N} h_i u_i f_i.$$

We see that if the right-hand side is zero, $f_i = 0$ for $i = 1, 2, \dots, N$, then u_i must be zero as well: For $i = 1$ and $i = N$, this is clearly true because of the first and last term on the left. If those are, however, zero, then the middle term containing the sum also implies that $u_2 = 0$ and $u_{N-1} = 0$, and the result follows by induction. Thus, our linear system (3.18) only has the identically zero solution when the right-hand side is zero, meaning that the system is nonsingular. As a result, the solution of (3.18) exists and is unique for any right-hand-side f.

We now focus on the error estimate, which is based on a similar summation formula for the error. We start with the integrated equation (3.17),

$$-\left(u_x(x_{i+\frac{1}{2}}) - u_x(x_{i-\frac{1}{2}})\right) = h_i f_i,$$

which implies using Taylor series that the exact solution satisfies

$$-\left(\frac{u(x_{i+1}) - u(x_i)}{h_{i+\frac{1}{2}}} - \frac{u(x_i) - u(x_{i-1})}{h_{i-\frac{1}{2}}}\right) = h_i f_i + R_{i-\frac{1}{2}} - R_{i+\frac{1}{2}},$$

where the remainder terms from the Taylor expansion satisfy $|R_{i\pm\frac{1}{2}}| \le Ch$. Subtracting from this equation the equation satisfied by the numerical approximation (3.18), we obtain that the errors satisfy the equation

$$\frac{e_{i+1} - e_i}{h_{i+\frac{1}{2}}} - \frac{e_i - e_{i-1}}{h_{i-\frac{1}{2}}} = R_{i-\frac{1}{2}} - R_{i+\frac{1}{2}}.$$

Multiplying on both sides by e_i and summing, we get

$$\sum_{i=1}^{N} \frac{(e_{i+1}-e_i)e_i}{h_{i+\frac{1}{2}}} - \sum_{i=1}^{N} \frac{(e_i - e_{i-1})e_i}{h_{i-\frac{1}{2}}} = \sum_{i=1}^{N} R_{i-\frac{1}{2}} e_i - \sum_{i=1}^{N} R_{i+\frac{1}{2}} e_i. \tag{3.23}$$

Using that $e_0 = e_{N+1} = 0$, we can add a vanishing first term to the first sum on the left and a vanishing last term to the second sum on the left and then sum them differently to obtain a sum of squares,

$$\frac{e_1e_0 - e_0^2}{h_{\frac{1}{2}}} + \frac{e_2e_1 - e_1^2}{h_{\frac{3}{2}}} + \frac{e_3e_2 - e_2^2}{h_{\frac{5}{2}}} + \cdots + \frac{e_Ne_{N-1} - e_{N-1}^2}{h_{N-\frac{1}{2}}} + \frac{e_{N+1}e_N - e_N^2}{h_{N+\frac{1}{2}}}$$
$$- \frac{e_1^2 - e_1e_0}{h_{\frac{1}{2}}} - \frac{e_2^2 - e_1e_2}{h_{\frac{3}{2}}} - \frac{e_3^2 - e_2e_3}{h_{\frac{5}{2}}} - \cdots - \frac{e_N^2 - e_Ne_{N-1}}{h_{N-\frac{1}{2}}} - \frac{e_{N+1}^2 - e_{N+1}e_N}{h_{N+\frac{1}{2}}}$$
$$= -\sum_{i=0}^{N} \frac{(e_{i+1} - e_i)^2}{h_{i+\frac{1}{2}}}.$$

Similarly, we proceed for the right-hand side, where we add a last vanishing term to the first sum and a first vanishing term to the second sum,

$$R_{\frac{1}{2}}e_1 + R_{\frac{3}{2}}e_2 + \cdots + R_{N-\frac{1}{2}}e_N + R_{N+\frac{1}{2}}e_{N+1}$$
$$-R_{\frac{1}{2}}e_0 + -R_{\frac{3}{2}}e_1 - \cdots - R_{N-\frac{1}{2}}e_{N-1} - R_{N+\frac{1}{2}}e_N$$
$$= \sum_{i=0}^{N} R_{i+\frac{1}{2}}(e_{i+1} - e_i).$$

Using the fact that all remainder terms $R_{i+\frac{1}{2}}$ are bounded in modulus by Ch, taking the modulus on both sides we thus obtain

$$\sum_{i=0}^{N} \frac{(e_{i+1} - e_i)^2}{h_{i+\frac{1}{2}}} \leq Ch \sum_{i=0}^{N} |e_{i+1} - e_i|. \tag{3.24}$$

We use now the Cauchy–Schwarz inequality for the term on the right,

$$\sum_{i=0}^{N} |e_{i+1} - e_i| = \sum_{i=0}^{N} \frac{|e_{i+1} - e_i|}{\sqrt{h_{i+\frac{1}{2}}}} \sqrt{h_{i+\frac{1}{2}}} \leq \sqrt{\sum_{i=0}^{N} \frac{(e_{i+1} - e_i)^2}{h_{i+\frac{1}{2}}}} \sqrt{\sum_{i=0}^{N} h_{i+\frac{1}{2}}}, \tag{3.25}$$

and because the sum of all mesh sizes $h_{i+\frac{1}{2}} = x_{i+1} - x_i$ satisfies $\sum_{i=0}^{N} h_{i+\frac{1}{2}} = 1$, by inserting (3.25) into (3.24) and dividing by a square root we obtain

$$\sqrt{\sum_{i=0}^{N} \frac{(e_{i+1} - e_i)^2}{h_{i+\frac{1}{2}}}} \leq Ch, \tag{3.26}$$

which is our first convergence estimate. To obtain the second estimate, we use the fact that e_i can be written as a telescopic sum using $e_0 = 0$,

$$e_i = \sum_{j=1}^{i} (e_j - e_{j-1}),$$

which implies, taking norms and inserting (3.26) into estimate (3.25), that

$$|e_i| \leq Ch.$$

□

The proof of Theorem 3.4 does not require the maximum principle, which was an essential ingredient in the convergence proof of the finite difference method in the previous chapter. In fact, Theorem 3.4 does not even require the consistency of the finite difference stencil obtained. It, however, uses the two major properties of finite volume methods:

1. The fluxes are conserved; i.e., the flux approximation used by the left cell of a cell boundary is the same as the one used by the right cell of the same cell boundary, which allowed us to combine the two separate sums in (3.23) on the right.

2. The flux approximation is consistent; i.e., the remainder terms $R_{i+\frac{1}{2}}$ become small when the mesh size becomes small.

For any finite difference scheme that satisfies these two properties, one can prove convergence using the finite volume techniques of Theorem 3.4.

The convergence result from Theorem 3.4 is, however, not sharp, and often quadratic convergence is observed. We can test this using the following simple MATLAB commands:

```
u=@(x) sin(pi*x);                                    % solution and corresponding
f=@(x) -pi^2*sin(pi*x);                              % right hand side
d=0.5;                                               % perturbation
NN=10*2.^(0:5);                                      % of the mesh, 0<=d<=0.5
err=[];
for j=1:length(NN)                                   % mesh refinement
  N=NN(j);
  if j==1
    h=1/N; xx=(0:h:1);
    xx(2:end-1)=xx(2:end-1)+(rand(1,N-1)-1/2)*d*h; % perturb primal mesh
    hhd=diff(xx);
    xxd=xx(1:end-1)+(1/2+(rand(1,N)-1/2)*d).*hhd;  % perturb dual mesh
    hh=diff([0 xxd 1]);
    figure(1);plot(xx,1,'o',xxd,1,'+');
  else
    xx(1:2:2*length(xx)-1)=xx;                       % refine mesh
    xx(2:2:end-1)=xxd;
    hhd=diff(xx);
    xxd=xx(1:end-1)+1/2.*hhd;
    hh=diff([0 xxd 1]);
  end;
  A=sparse(N-1);
  for i=1:N-1                                        % assemble matrix by going
    A(i,i)=-(1/hhd(i)+1/hhd(i+1))/hh(i+1);           % over all cell interfaces
    if i>1, A(i,i-1)=1/hhd(i)/hh(i+1); end;
    if i<N-1, A(i,i+1)=1/hhd(i+1)/hh(i+1); end;
  end;
  b=f(xx(2:end-1)'); ue=u(xx');
  ua=A\b;
  err(j)=max(abs(ue(2:end-1)-ua))
end
figure(2);loglog(1./NN,err,'--',1./NN,1./NN,'-',1./NN,1./NN.^2,'-')
```

We show in Figure 3.12 on the left the random initial mesh used for the experiment and on the right the error measured in the maximum norm when the mesh is regularly refined several times. Clearly, the finite volume method converges quadratically, even though our convergence estimate in Theorem 3.4 shows only linear convergence. It is an open problem to prove this often observed quadratic convergence for the general case of finite volume methods [17]. In specific situations it is, however, possible; see, for example, Forsythe and Sammon [18], who prove quadratic convergence for cell-centered

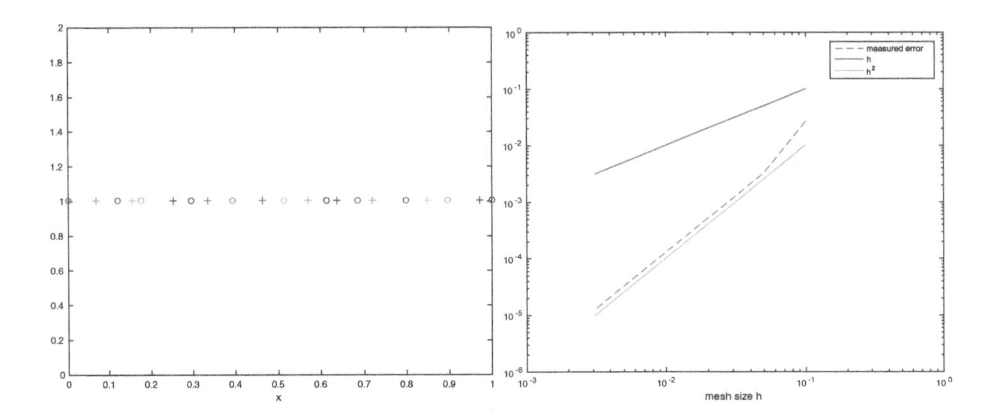

Figure 3.12. *Left: Initial randomly perturbed finite volume mesh leading to inconsistent finite difference stencils when the finite volume method is applied. Right: Convergence measured in the maximum norm when the mesh is refined in a regular way and reference lines for linear and quadratic convergence.*

rectangular nonuniform grids and also show that lower-order boundary condition discretizations do not cause any harm.

3.6 ▪ Concluding Remarks

Over the last two decades finite volume methods have been developed much further than what we have seen in this chapter: An excellent introduction can be found in [17]. In particular, the discrete duality finite volume method has now so much mathematical structure that fully general convergence proofs on arbitrary domains with arbitrary meshes have become possible; see, for example, [12, 1]. These methods are also very much related to a new class of finite element methods, namely, the discontinuous Galerkin methods. Finite volume methods have the following advantages and disadvantages, denoted by plus and minus signs, respectively:

+ The finite volume method gives a systematic way to obtain discretization stencils for PDEs.

+ The finite volume method works for arbitrary geometries and meshes.

– The method is harder to implement when on general meshes, and one has to follow an assembly strategy as in the finite element method, which we will see later.

– Higher-order methods require sophisticated flux definitions.

3.7 ▪ Problems

Problem 3.1 (finite volumes and finite differences).

1. Construct a finite volume discretization of the nonlinear equation

$$\begin{cases} (uu_x)_x &= f(x), \\ u(0) &= 0, \\ u(1) &= 0. \end{cases}$$

Hint: To complete the scheme, approximate the unknown function u at the midpoints using an average of the neighboring values.

2. Show that the truncation error is $O(h^2)$.

3. Show that a reformulation of the differential equation allows you to obtain the same scheme using centered finite differences.

4. Implement the finite volume method you obtained, and use Newton's method to solve the nonlinear system of equations you obtain. Show graphically that the method converges. Is the solution unique?

Problem 3.2 (nonconstant diffusion coefficients). We consider the two-dimensional problem

$$\nabla \cdot (a(x, y)\nabla u(x, y)) = f(x, y), \tag{3.27}$$

where a is a scalar function. We assume that the grid for the finite volume method to be constructed is rectangular and regular.

1. Derive a finite volume discretization of (3.27).

2. Show that the scheme you obtained in the one-dimensional case has a second-order truncation error.

3. Using Maple, show that the two-dimensional scheme is also second order.

Problem 3.3 (Neumann boundary conditions for finite volumes).

1. Consider the Poisson equation $-\Delta u = f$ discretized by a finite volume method on a regular rectangular grid, and discretize the Neumann boundary conditions using the *vertex-centered* approach. Show that the truncation error is of order one.

2. Show how the associated dicretization matrix can easily be modified to include the Neumann conditions discretized with the *vertex-centered* approach.

3. Repeat parts 1 and 2 for a *cell-centered* discretization of the Neumann conditions.

4. What relation can you find between the finite difference and finite volume discretizations of the Neumann conditions in this case?

Problem 3.4 (consistency of finite volumes). Use the Maple program statements in this chapter to explore under which conditions the FV4 scheme is consistent. Can you also find conditions when one uses four neighboring cells, except for perfect symmetry?

Chapter 4
The Spectral Method

Sehr viel besser eignen sich Entwicklungen nach Polynomen, Fouriersche Reihen usw. für die Darstellung einer reellen Funktion $w(x, y, \ldots)$ in einem gegebenen Bereich.[a]

Walther Ritz, Über eine neue Methode zur Lösung gewisser Variationsprobleme der mathematischen Physik, 1908

The purpose of the following article is to show the manifold possibilities of the application of Tshebysheff's polynomials in approximation problems. All these applications are based on some simple basic properties of the Fourier series. Although the mathematical nature of the Fourier series is exhaustively investigated, it is not sufficiently realized how excellent approximations, for both empirical and analytical functions, may be obtained by combining the advantages of the power series and the Fourier series.

Cornelius Lanczos, Trigonometric Interpolation of Empirical and Analytical Functions, 1938

Historically, second order accurate difference methods have been used for computations in dynamic meteorology and oceanography. We investigate more accurate difference methods and show that fourth order methods are optimal in some sense. This method is then compared with a variant of the Fourier technique.

Heinz-Otto Kreiss and Joseph Oliger, Comparison of accurate methods for the integration of hyperbolic equations, 1972

[a] Expansions in polynomials, Fourier series, etc. are much more suitable for the representation of a real function $w(x, y, \ldots)$ in a given domain.

We now introduce a very different approach for solving PDEs, namely, the *spectral methods*. The idea behind these methods is to search for a solution as a series generated by a set of basis functions, i.e.,

$$u(\mathbf{x}) = \sum_{k=-\infty}^{\infty} \hat{u}(k)\varphi_k(\mathbf{x}), \tag{4.1}$$

and thus goes back to the invention of Fourier and separation of variables [19], as we have seen in Chapter 1. As part of his groundbreaking work on vibrating plates in 1908, Ritz was the first to propose using a truncated expansion of this form as a computational method: In order to approximate the shape of a two-dimensional bending or vibrating plate, he used basis functions φ_k that are products of eigenfunctions of the one-dimensional bar; see the quote above and Figure 4.1, extracted from [49]; see also [50]. Ritz could thus be regarded as the father of spectral methods.

Es wird unten gezeigt werden, wie bei gegebener Berandung die bekannten Eigenschaften der Polynome, Fourier-Reihen usw. es gestatten, solche Funktionen ψ_i zu bilden.

Um nun die gesuchten sukzessiven Approximationen zu erhalten, haben wir nur den Ausdruck

$$w_m = a_1\psi_1 + a_2\psi_2 + \ldots + a_m\psi_m$$

an Stelle von w *in das Integral* J *zu setzen;* sei

$$\text{(8)} \qquad J_m = \int\int_{\text{D}} \left[\frac{1}{2}(\Delta w_m)^2 - f w_m\right] dS,$$

Dagegen führt die physikalische Analogie von selbst zu Funktionen, die die Eigenschaften der ψ_i besitzen : die Betrachtung eines an beiden Enden eingeklemmten, elastischen Stabes, gewissermassen das Analogon zum vorgelegten Problem in *einer* Dimension, wird uns die gewünschten ψ_i in zweckmässiger Form liefern.

Figure 4.1. *Groundbreaking proposition of Ritz from* 1908 *to use a truncated expansion of the form* (4.1) *in minimizing an energy integral whose stationary point is a solution to the biharmonic equation, using products of solutions of the one-dimensional problem.*

In modern spectral methods, the functions φ_k are generally chosen from one of the following two categories:

- trigonometric functions,
- orthogonal polynomials.

Lanczos already noticed in 1938 that important approximation advantages can be gained by considering polynomial approximations in a similar sense as Fourier series, namely, globally, as opposed to Taylor series [35]. Spectral methods, in contrast to all the methods we have seen so far, are global in nature, and it is natural to start this chapter with Fourier series.

4.1 ▪ Spectral Method Based on Fourier Series

Suppose we are interested in computing an approximate solution of the one-dimensional Poisson equation with periodic boundary conditions,

$$\left\{\begin{array}{rcll} u_{xx} & = & f, & \Omega = (0, 2\pi), \\ u(0) & = & u(2\pi), & \\ u'(0) & = & u'(2\pi). & \end{array}\right. \tag{4.2}$$

A periodic function $u \in \mathcal{C}(0, 2\pi)$ admits a Fourier series,

$$u(x) = \sum_{k=-\infty}^{\infty} u_k e^{ikx}, \tag{4.3}$$

where the Fourier coefficients are given by the formula

$$u_k = \frac{1}{2\pi}\int_0^{2\pi} u(x)e^{-ikx}dx. \tag{4.4}$$

In order to solve the Poisson equation (4.2), one can insert the Fourier series representation of the solution (4.3) into the equation to obtain

$$\partial_{xx}\left(\sum_{k=-\infty}^{\infty} u_k e^{ikx}\right) = \sum_{k=-\infty}^{\infty} -k^2 u_k e^{ikx} = f. \tag{4.5}$$

To determine the unknown Fourier coefficients u_k, a natural idea is to use the orthogonality of the exponential functions e^{ikx}: We multiply both sides by the "test function"[16] e^{-ilx} and integrate over $(0, 2\pi)$, which leads to

$$\sum_{k=-\infty}^{\infty} -k^2 u_k \int_0^{2\pi} e^{ikx}e^{-ilx}dx = \int_0^{2\pi} f(x)e^{-ilx}dx = 2\pi f_l, \tag{4.6}$$

where f_l is the lth Fourier coefficient of $f(x)$. Using now the orthogonality of e^{ikx}, i.e.,

$$\int_0^{2\pi} e^{ikx}e^{-ilx}dx = \begin{cases} 0 & \text{if } l \neq k, \\ 2\pi & \text{if } l = k, \end{cases}$$

we obtain a very simple equation for the Fourier coefficients u_k, namely,

$$-l^2 u_l = f_l. \tag{4.7}$$

The Fourier coefficients of the solution are therefore given by

$$u_l = \begin{cases} -\frac{1}{l^2} f_l & \text{if } l \neq 0, \\ \text{undefined} & \text{if } l = 0. \end{cases} \tag{4.8}$$

Remark 4.1. *If* $l = 0$, (4.7) *becomes*

$$0 = f_0 = \frac{1}{2\pi}\int_0^{2\pi} f(x)dx.$$

Therefore, (4.2) *has a solution if and only if f has a vanishing mean. This condition is known as a* compatibility condition. *Note that the compatibility condition does not imply that the solution is unique because u_0 is undefined and can be chosen freely. To get a unique solution, we need to specify the value of u_0; see also Problem* 4.1.

Once Fourier coefficients are found, the solution can be written as

$$u(x) = \hat{u}(0) - \sum_{k\neq 0} \frac{1}{k^2} f_k e^{ikx}.$$

To obtain a numerical approximation, the infinite expansion must be truncated, as Ritz proposed, so we need to estimate the error induced by such a truncation. This error

[16]This idea will appear again in the next chapter on the finite element method, when the basis functions are no longer orthogonal. Indeed, it is precisely this approach for a sine expansion that motivated Galerkin in 1915 to proceed in the same way with more general functions φ; see [20, pp. 169–171].

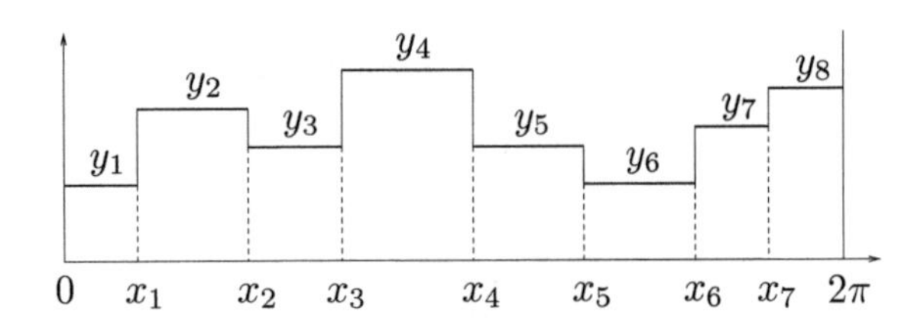

Figure 4.2. *Example of a step function in the proof of Theorem* 4.2.

estimate is substantially different from error estimates we have seen so far; see the last quote at the beginning of this chapter from [34], where the authors point out important qualities, but also limitations, of the Fourier spectral methods.[17]

To study rigorously the truncation error, we need to study the decay of the Fourier coefficients u_k. We do this for quite a general class of functions.

Definition 4.1 (functions of bounded variation). *For a function* $u : [0, 2\pi] \to \mathbb{R}$, *its* total variation *is defined by*

$$V_{[0,2\pi]}u := \sup_{n>0} \sup_{0=x_0<x_1<\cdots<x_n=2\pi} \left(\sum_{i=0}^{n-1} |u(x_{i+1}) - u(x_i)| \right).$$

The function u *is of* bounded variation *if* $V_{[0,\pi]}u < \infty$.

If u is of bounded variation, u is integrable in the sense of Riemann [29], and continuously differentiable functions are of bounded variation. Continuity alone is, however, not sufficient for bounded variation; e.g., $u(x) = x\sin(1/x)$ is not of bounded variation, but there are also functions that are not continuous but of bounded variation, for example, the step functions. We can now prove the following essential estimate, which leads to the spectacular so-called *spectral convergence* of spectral methods.

Theorem 4.2. *If* $u : [0, 2\pi] \mapsto \mathbb{R}$ *is of bounded variation, then*

$$|u_k| \leq \frac{C}{|k|} \quad \text{for some } C > 0 \text{ independent of } k. \tag{4.9}$$

If $u : \mathbb{R} \mapsto \mathbb{R}$ *is* 2π*-periodic and* p *times differentiable with* $u^{(p)}|_{[0,2\pi]}$ *of bounded variation, then there exists a constant* $C > 0$ *depending on* $u^{(p)}$ *but independent of* k *such that*

$$|u_k| \leq \frac{C}{|k|^{p+1}}. \tag{4.10}$$

Proof. We start by showing (4.9) for a step function $u(x)$, as illustrated in Figure 4.2. For such a step function $u(x)$, we can explicitly compute its Fourier coefficients,

$$\begin{aligned} u_k &= \frac{1}{2\pi}\int_0^{2\pi} u(x)e^{-ikx}dx = \frac{1}{2\pi}\sum_{j=1}^{n} y_j \int_{x_{j-1}}^{x_j} e^{-ikx}dx \\ &= \frac{1}{2\pi ik}\left(y_1 + e^{ikx_1}(y_2 - y_1) + e^{ikx_2}(y_3 - y_2) + \cdots - y_n\right), \end{aligned}$$

[17] "It is at the present time not clear what the accuracy of the Fourier method is for equations with variable coefficients, particularly when discontinuities are present. Some preliminary calculations have shown that if the solution is discontinuous, then the number of necessary frequencies must be increased substantially."

and therefore we obtain the desired estimate

$$|u_k| \leq \frac{1}{2\pi|k|}\left(V_{[0,2\pi]}u + |u(2\pi) - u(0)|\right) \leq \frac{C}{|k|}.$$

Let $u(x)$ now be an arbitrary function of bounded variation. Since such a $u(x)$ is integrable, by the definition of the Riemann integral, there exists a step function $\tilde{u}(x)$ which satisfies $V_{[0,2\pi]}\tilde{u} + |\tilde{u}(2\pi) - \tilde{u}(0)| \leq V_{[0,2\pi]}u + |u(2\pi) - u(0)|$ such that $\int_0^{2\pi} \tilde{u}(x)e^{-ikx}dx$ is arbitrarily close to the Fourier coefficient $\int_0^{2\pi} u(x)e^{-ikx}dx$ of $u(x)$, which proves (4.9).

Now if $u(x)$ is once differentiable with its derivative of bounded variation, we can use integration by parts,

$$u_k = \frac{1}{2\pi}\int_0^{2\pi} u(x)e^{-ikx}dx = \frac{1}{2\pi}u(x)\frac{e^{-ikx}}{-ik}\bigg|_0^{2\pi} + \frac{1}{2\pi ik}\int_0^{2\pi} u'(x)e^{-ikx}dx,$$

and since the first term on the right-hand side vanishes because of periodicity, we can apply (4.9) to $u'(x)$ to get $|u_k| \leq \frac{C}{|k|^2}$. Using integration by parts several times then leads to the estimate (4.10). □

We show in Figure 4.3 for several functions of increasing regularity how their Fourier coefficients decay. The figures were obtained with the following MATLAB code:

```
f{1}=inline('exp(x)-(exp(2*pi)-1)/2/pi','x');      % u discontinuous at 0 -> h
f{2}=inline('2/3*pi^2-2*pi*x+x.^2','x');            % u continuous at 0    -> h^2
f{3}=inline('2*pi^2*x-3*pi*x.^2+x.^3','x');         % u' continuous at 0   -> h^3
f{4}=inline('-8/15*pi^4+4*pi^2*x.^2-4*pi*x.^3+x.^4','x');  % u'' cont  -> h^4
f{5}=inline('-8/3*pi^4*x+20/3*pi^2*x.^3-5*pi*x.^4+x.^5','x'); % u'''cont->h^5
f{6}=inline('3./(5-4*cos(x))-1','x');               % all derivatives continuous
n=512; h=2*pi/n; x=(0:h:2*pi)';
for i=1:6
  b=feval(f{i},x(1:n)); bh=fft(b);
  subplot(2,1,1); plot([x;x+2*pi],f{i}([x;x]),'-'); xlabel('x');
  subplot(2,1,2);
  if i<6
    loglog(1:n/2-1,abs(bh(2:n/2)),'o',1:n/2-1,abs(bh(2))./(1:n/2-1).^i,'-');
    xlabel('k'); legend('u_k',['1/k^' num2str(i)] );
  else
    loglog(1:n/2-1,abs(bh(2:n/2)),'o');
    xlabel('k'); legend('u_k');
  end
  pause
end;
```

Note that we already use a very efficient way to compute the Fourier coefficients, the fast Fourier transform, which we will investigate more closely in the next section. With Theorem 4.2, we can now precisely estimate the difference between a truncated Fourier series expansion of $u(x)$ and the function $u(x)$ itself.

Theorem 4.3. *Let $u : \mathbb{R} \mapsto \mathbb{R}$ be 2π-periodic and $p \geq 1$ times differentiable with $u^{(p)}|_{[0,2\pi]}$ of bounded variation. For an even number $n > 0$, let*

$$u_h(x) := \sum_{|k|\leq n/2} u_k e^{ikx}$$

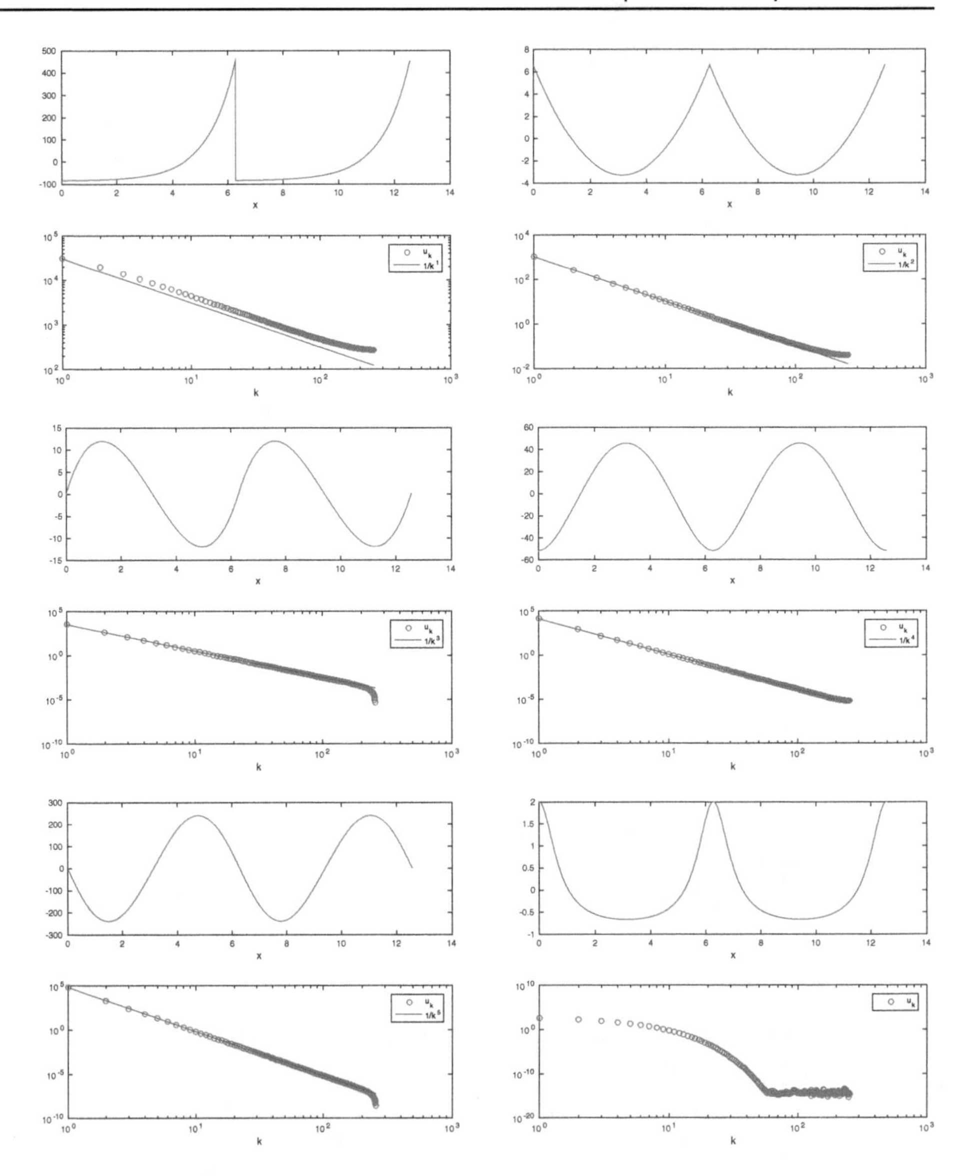

Figure 4.3. *Illustration of how the Fourier coefficients decay faster and faster if the function becomes more and more regular. From top left to bottom right: Functions with zero up to four derivatives of bounded variation and a function with all derivatives continuous in the last panel, where one clearly sees spectral convergence.*

be the truncated Fourier expansion of $u(x)$*. Then we have the error estimate*

$$||u - u_h||_2 \leq \frac{2^{p+1}\sqrt{2\pi}C}{\sqrt{2p+1}}\frac{1}{n^{p+\frac{1}{2}}}, \tag{4.11}$$

where $C > 0$ *is a constant independent of* n*.*

Proof. The proof is based on the Parseval–Plancherel identity,

$$||u||_2^2 = 2\pi \sum_{k=-\infty}^{\infty} |u_k|^2,$$

which implies, using Theorem 4.2, that

$$||u-u_h||_2^2 = 2\pi \sum_{|k|>n/2} |u_k|^2 \leq 2\pi \sum_{|k|>n/2} \frac{C^2}{|k|^{2p+2}} \leq 4\pi C^2 \int_{n/2}^{\infty} \frac{1}{k^{2p+2}} dk$$

$$= \frac{2^{2p+3}\pi C^2}{2p+1} \frac{1}{n^{2p+1}}.$$

□

How should we interpret this result in order to compare it to error estimates we have obtained for the finite difference and finite volume methods? If we assume that the unknowns are the Fourier coefficients u_k, then we have $n+1$ unknowns on the interval $[0, 2\pi]$, which represents the problem domain, and thus we can define a hypothetical mesh size[18] $h := \frac{2\pi}{n}$ and rewrite the error estimate (4.11) as a function of this mesh size h,

$$||u-u_h||_2 \leq \frac{2C}{\sqrt{2p+1}} \frac{1}{\pi^p} h^{p+\frac{1}{2}} = O(h^{p+\frac{1}{2}}).$$

This reveals the most important property of spectral methods: They converge at a rate depending on the regularity p of the solution. If the solution has two derivatives, convergence will be $O(h^{2.5})$; if it has 10, it will be $O(h^{10.5})$; and if it has an infinite number of derivatives, it converges faster than $O(h^p)$ for any $p > 1$. This is called spectral convergence, a convergence that is faster than any algebraic rate in the grid size h.

Our analysis so far has been for a continuous solution $u(x)$, and we only truncated the Fourier series approximation, without solving the PDE. To obtain a real numerical method, we need to approximate also the integral in the evaluation of the Fourier coefficient and solve the PDE, which will lead to a fully discrete version of the Fourier spectral method.

4.2 ▪ Spectral Method with Discrete Fourier Series

Suppose u is a 2π-periodic function. Just as for finite difference and finite volume methods, we now introduce a mesh with mesh points

$$x_j = jh, \quad h = \frac{2\pi}{n} \quad \text{for some even } n, \tag{4.12}$$

and assume that we only have knowledge of the function $u(x)$ at the grid points x_j. Recalling the formula of the Fourier series,

$$u(x) = \sum_{k=-\infty}^{\infty} u_k e^{ikx}, \tag{4.13}$$

we see that we cannot possibly determine $u(x)$ uniquely since there are infinitely many coefficients u_k but only a finite number of function values $\{u(x_j)\}_{j=0}^{n-1}$ at the grid points. This *interpolation problem* has infinitely many solutions, one of which is of the form

$$u(x_j) = \sum_{k=-n/2}^{n/2-1} \hat{u}_k e^{ikx_j}, \quad j = 0, \ldots, n-1. \tag{4.14}$$

[18]This will become a real mesh size soon.

In other words, we can obtain one such interpolant by setting all the Fourier coefficients outside the range $k \in K = \{-\frac{n}{2}, \dots, \frac{n}{2} - 1\}$ to zero and calculating the remaining n Fourier coefficients from the n grid point equations. The resulting trigonometric interpolant, $\hat{u}(x) := \sum_{k=-n/2}^{n/2-1} \hat{u}_k e^{ikx}$, coincides with $u(x)$ at the interpolation points x_j but not necessarily anywhere else. Thus, we generally have $u(x) \neq \hat{u}(x)$ or, equivalently, $u_k \neq \hat{u}_k$. Nonetheless, there is a strong relation between u_k and $\hat{u}_k$, which we will prove later in Theorem 4.6.

To find an explicit formula for the $\hat{u}_k$, it is instructive to rewrite the finite expansion (4.14) in matrix form. We denote

$$
\begin{aligned}
\mathbf{u} &:= (u(x_0), u(x_1), \dots, u(x_{n-1}))^T, \\
\hat{\mathbf{u}} &:= (\hat{u}_{-\frac{n}{2}}, \hat{u}_{-\frac{n}{2}+1}, \dots, \hat{u}_{\frac{n}{2}-1})^T, \\
F^{-1} &:= \begin{pmatrix}
e^{i\frac{-n}{2}x_0} & e^{i(\frac{-n}{2}+1)x_0} & \dots & e^{i(\frac{n}{2}-1)x_0} \\
e^{i\frac{-n}{2}x_1} & e^{i(\frac{-n}{2}+1)x_1} & \dots & e^{i(\frac{n}{2}-1)x_1} \\
\vdots & & & \vdots \\
e^{i\frac{-n}{2}x_{n-1}} & e^{i(\frac{-n}{2}+1)x_{n-1}} & \dots & e^{i(\frac{n}{2}-1)x_{n-1}}
\end{pmatrix},
\end{aligned}
$$

where F^{-1} is invertible, with an inverse F that will be defined soon. We can then rewrite the expansion (4.14) in matrix form,

$$
\mathbf{u} = F^{-1}\hat{\mathbf{u}}, \tag{4.15}
$$

and by multiplying by F on both sides, we get

$$
\hat{\mathbf{u}} = F\mathbf{u}. \tag{4.16}
$$

The matrix F is called the *discrete Fourier transform* (DFT) and F^{-1} the *inverse discrete Fourier transform* (IDFT). To find the discrete Fourier coefficients $\hat{u}_k$ from the function values $u(x_j)$, we first notice the discrete orthogonality relation

$$
\begin{aligned}
\sum_{j=-n/2}^{n/2-1} e^{ikx_j} e^{-ilx_j} &= \sum_{j=-n/2}^{n/2-1} e^{i(k-l)\frac{2\pi}{n}j} = e^{-i(k-l)\pi} \sum_{j=0}^{n-1} e^{i(k-l)\frac{2\pi}{n}j} \\
&= \begin{cases} e^{-i(k-l)\pi} \frac{1-e^{i(k-l)2\pi}}{1-e^{i(k-l)\frac{2\pi}{n}}} = 0 & \text{if } k \neq l \mod n, \\ n & \text{if } k = l \mod n, \end{cases}
\end{aligned} \tag{4.17}
$$

where we used the summation formula for geometric series. In analogy with the continuous formula for the Fourier coefficients (4.4), we multiply (4.14) by e^{-ilx_j} and sum over $j = -\frac{n}{2}, \dots, \frac{n}{2} - 1$ to obtain

$$
\sum_{j=-n/2}^{n/2-1} u(x_j)e^{-ilx_j} = \sum_{j=-n/2}^{n/2-1} \sum_{k=-n/2}^{n/2-1} \hat{u}_k e^{i(k-l)x_j} = n\hat{u}_l,
$$

and therefore the explicit expression for the DFT (and thus the matrix F) is

$$
\hat{u}_k = \frac{1}{n} \sum_{j=-n/2}^{n/2-1} u(x_j)e^{-ikx_j}. \tag{4.18}
$$

Remark 4.2. *We can also find the matrix F by applying the trapezoidal quadrature rule to approximate the integral in the definition of the Fourier coefficient,*

$$\hat{u}_k = \frac{1}{n}\sum_{j=-n/2}^{n/2-1} u(x_j)e^{-ikx_j} \approx \frac{1}{2\pi}\int_{-\pi}^{\pi} u(x)e^{-ikx}dx = u_k.$$

Imposing the Fourier series expansion (4.14) *at discrete mesh points is therefore equivalent to approximating the Fourier coefficient integral by quadrature.*

We need a final ingredient in order to be able to give our first spectral method.

Definition 4.4 (differentiation matrix). *A* differentiation matrix *is a matrix D that, when multiplied with a vector $\mathbf{u}$, gives an approximation of the derivative of the function $u(x)$ from which the vector $\mathbf{u}$ is sampled.*

We have seen many differentiation matrices so far; for example, (2.11) is a second-order differentiation matrix in two spatial dimensions, or

$$D^+ := \frac{1}{h}\begin{bmatrix} -1 & 1 & & \\ & \ddots & \ddots & \\ & & -1 & 1 \\ & & & -1 \end{bmatrix}, \quad D^- := \frac{1}{h}\begin{bmatrix} 1 & & & \\ -1 & 1 & & \\ & \ddots & \ddots & \\ & & -1 & 1 \end{bmatrix}$$

are the forward and backward finite difference differentiation matrices. For spectral methods, the differentiation matrices are defined using the Fourier transform. In analogy to the continuous situation, where a derivative in Fourier space corresponds to multiplication by ik as in (4.5), we define

$$\hat{D}_F := \begin{pmatrix} -i(\frac{n}{2}) & & & & & & \\ & -i(\frac{n}{2}-1) & & & & & \\ & & \ddots & & & & \\ & & & -i & & & \\ & & & & 0 & & \\ & & & & & i & \\ & & & & & & \ddots \\ & & & & & & & i(\frac{n}{2}-1) \end{pmatrix}, \tag{4.19}$$

and thus the Fourier differentiation matrix is obtained by first transforming the vector into the Fourier domain, applying $\hat{D}_F$, and then performing the inverse Fourier transform,

$$D_F := F^{-1}\hat{D}_F F. \tag{4.20}$$

This leads to the following *discrete Fourier spectral method* for solving the Poisson equation with periodic boundary conditions (4.2):

1. Discretize the right-hand side f by sampling,

$$\mathbf{f} := (f(x_0), \ldots, f(x_{n-1}))^T.$$

2. To solve the discretized version of $u_{xx} = f$ using the Fourier spectral method, we have to solve

$$D_F^2 \mathbf{u} = \mathbf{f} \quad \Longrightarrow \quad \mathbf{u} = D_F^{-2}\mathbf{f}.$$

We thus have to apply

$$D_F^{-2} = (F^{-1}\hat{D}_F F)^{-2} = (F^{-1}\hat{D}_F^{-1}F)(F^{-1}\hat{D}_F^{-1}F) = F^{-1}\hat{D}_F^{-2}F$$

to the right-hand side $\mathbf{f}$, which requires the following steps:

(a) Fourier transform $\mathbf{f}$ to get $\hat{\mathbf{f}}$;

(b) apply $\hat{D}_F^{-2}$ to $\hat{\mathbf{f}}$;

(c) back transform the result to obtain $\mathbf{u}_h := F^{-1}\hat{D}_F^{-2}F\mathbf{f}$ as an approximation of

$$\mathbf{u} := (u(x_0), \ldots, u(x_{n-1}))^T.$$

As in the continuous case, we must be careful when applying $\hat{D}_F^{-2}$ since this matrix is singular; see the zero in the center in (4.19). It is at this moment that the ill-posedness of the problem manifests itself and a choice has to be made; see Remark 4.1.

Remark 4.3. *To obtain an efficient algorithm, a key ingredient is to use the* fast Fourier transform *(FFT) instead of multiplying with the matrix F and the* inverse fast Fourier transform *(IFFT) instead of multiplying with the matrix F^{-1}, as we have done already in the MATLAB script above. In MATLAB, the truncated discrete Fourier series is defined as*

$$u(x_j) = \frac{1}{n}\sum_{k=0}^{n-1} \tilde{u}_k e^{ikx_j} \tag{4.21}$$

instead of our choice shown in (4.14)*. The next lemma shows that there is a direct link between the coefficients $\tilde{u}_k$ from MATLAB and the coefficients $\hat{u}_k$ we used.*

Lemma 4.5. *The coefficients $\tilde{u}_k$ and $\hat{u}_k$ for $l = 0, 1, \ldots, \frac{n}{2} - 1$ are related by*

$$\hat{u}_{-\frac{n}{2}+l} = \frac{1}{n}\tilde{u}_{\frac{n}{2}+l}, \quad \hat{u}_l = \frac{1}{n}\tilde{u}_l.$$

Proof. With $k := -\frac{n}{2} + l$ for $l = 0, 1, \ldots, \frac{n}{2} - 1$, and using that the exponential function is 2π-periodic and the definition of x_j, we obtain

$$e^{ikx_j} = e^{i(-\frac{n}{2}+l)j\frac{2\pi}{n}} = e^{i(-j\pi)+lj\frac{2\pi}{n}} = e^{i(j\pi+lj\frac{2\pi}{n})} = e^{i(\frac{n}{2}+l)j\frac{2\pi}{n}} = e^{i\tilde{k}x_j},$$

where $\tilde{k} = \frac{n}{2} + l$. Thus, the coefficients of the sum (4.21) are linked to the coefficients of (4.14) by replacing the indices $k = -\frac{n}{2} + l$ by $\tilde{k} = \frac{n}{2} + l$ for $l = 0, 1, \ldots, \frac{n}{2} - 1$ and leaving them unchanged for $k = 0, 1, \ldots, \frac{n}{2} - 1$. □

If we want to use the FFT/IFFT in MATLAB, the differentiation matrix in (4.19) should thus be reordered,

$$\hat{D}_F = \begin{pmatrix} 0 & & & & & & & & \\ & i & & & & & & & \\ & & 2i & & & & & & \\ & & & \ddots & & & & & \\ & & & & (\frac{n}{2}-1)i & & & & \\ & & & & & \frac{n}{2}i & & & \\ & & & & & & -(\frac{n}{2}-1)i & & \\ & & & & & & & \ddots & \\ & & & & & & & & -i \end{pmatrix}.$$

4.3 ▪ Convergence Analysis

The two sources of error in the fully discrete spectral method described above are as follows:

1. quadrature: $u_k \overset{DFT}{\rightarrow} \hat{u}_k$;
2. truncation: $\sum_{k\in\mathbb{Z}} u_k \rightarrow \sum_{-n/2}^{n/2-1} \hat{u}_k$.

We have already studied the error induced by the truncation of the Fourier series at the continuous level; see Theorem 4.2. However, we have not yet considered the quadrature error that results from using grid point values to compute $\hat{u}_k$ rather than evaluating the integral (4.4) exactly. The next result quantifies this error.

Theorem 4.6 (aliasing). *Let u_k be the Fourier series coefficients defined by (4.4) of a 2π-periodic function u, and let $\hat{u}_k$ be the discrete Fourier coefficients defined by (4.16). If the series $\sum_{k\in\mathbb{Z}} u_k$ is absolutely convergent, then*

$$\hat{u}_k - u_k = \sum_{\substack{l\in\mathbb{Z}\\ l\neq 0}} u_{k+ln}. \tag{4.22}$$

Proof. Starting with the definition of the discrete Fourier transform (4.18), we substitute the Fourier series representation of $u(x_j)$ from (4.13) to get

$$\hat{u}_k = \frac{1}{n}\sum_{j=-n/2}^{n/2-1} u(x_j)e^{-ikx_j} = \frac{1}{n}\sum_{j=-n/2}^{n/2-1}\sum_{l=-\infty}^{\infty} u_l e^{ilx_j}e^{-ikx_j}.$$

We now switch the order of summation, which is possible because the series is absolutely convergent, and apply the discrete orthogonality relation (4.17) to conclude that

$$\hat{u}_k = \frac{1}{n}\sum_{l=-\infty}^{\infty} u_l \sum_{j=-n/2}^{n/2-1} e^{i(l-k)x_j} = \sum_{l=-\infty}^{\infty} u_{k+ln},$$

from which we obtain the required result after subtracting u_k from both sides. ☐

With Theorems 4.2 and 4.6, we can now give a precise estimate of the error of the discrete Fourier spectral method.

Theorem 4.7 (spectral convergence). *Let $f : [0, 2\pi] \mapsto \mathbb{R}$ be 2π-periodic and $p \geq 1$ times differentiable with $f^{(p)}|_{[0,2\pi]}$ of bounded variation, and let $\mathbf{u}$ be samples of the exact solution of* (4.2) *on the grid* (4.12) *with $n \geq 4$ even. If $\mathbf{u}_h$ denotes the approximate solution obtained by the discrete Fourier spectral method, then it satisfies the error estimate*

$$||\mathbf{u} - \mathbf{u}_h||_2 \leq \frac{C}{n^{p+1}}, \qquad C \text{ some constant.} \tag{4.23}$$

Proof. To simplify the notation, we also define the set of frequencies used in the truncation $K := \{-\frac{n}{2}, -\frac{n}{2}+1, \dots, \frac{n}{2}-1\}$. Using the Parseval–Plancherel identity implies

$$\begin{aligned}
\|\mathbf{u} - \mathbf{u}_h\|_{L^2}^2 = 2\pi \sum_{k\in\mathbb{Z}} |u_k - \hat{u}_k|^2 \quad &= \quad 2\pi \left(\sum_{k\in K} |u_k - \hat{u}_k|^2 + \sum_{k\notin K} |u_k|^2 \right) \\
&= 2\pi \left(\sum_{k\in K\backslash 0} \frac{1}{k^4} |f_k - \hat{f}_k|^2 + \sum_{k\notin K} \frac{1}{k^4} |f_k|^2 \right),
\end{aligned} \tag{4.24}$$

where we used in the last step (4.7) and the differentiation matrix (4.19) from step 2(b) of the discrete Fourier spectral method. We now estimate the first sum in (4.24) using Theorems 4.6 and 4.2:

$$\begin{aligned}
\sum_{k\in K\backslash 0} \frac{1}{k^4} |f_k - \hat{f}_k|^2 = \sum_{k\in K\backslash 0} \frac{1}{k^4} \Big| \sum_{j\neq 0} f_{k+jn} \Big|^2 &\leq \sum_{k\in K\backslash 0} \frac{1}{k^4} \left(\sum_{j\neq 0} \frac{C}{|k+jn|^{p+1}} \right)^2 \\
&\leq \sum_{k=1}^{\frac{n}{2}-1} \frac{C^2}{k^4} \left(\sum_{j\geq 1} \frac{1}{|k+jn|^{p+1}} + \frac{1}{|k-jn|^{p+1}} \right)^2 \\
&\quad + \sum_{k=-\frac{n}{2}}^{-1} \frac{C^2}{k^4} \left(\sum_{j\geq 1} \frac{1}{|k+jn|^{p+1}} + \frac{1}{|k-jn|^{p+1}} \right)^2 .
\end{aligned} \tag{4.25}$$

Now the first double sum in (4.25) can be estimated by

$$\begin{aligned}
&\frac{C^2}{n^{2p+2}} \sum_{k=1}^{\frac{n}{2}-1} \frac{1}{k^4} \left(\sum_{j\geq 1} \frac{1}{|\frac{k}{n}+j|^{p+1}} + \frac{1}{|\frac{k}{n}-j|^{p+1}} \right)^2 \\
&\leq \frac{C^2}{n^{2p+2}} \sum_{k=1}^{\frac{n}{2}-1} \frac{1}{k^4} \left(\sum_{j\geq 1} \frac{1}{j^{p+1}} + \frac{1}{|j-\frac{1}{2}|^{p+1}} \right)^2 ,
\end{aligned}$$

where we replaced the denominators by their lower bounds, given the range of k. Now since both infinite sums in j converge for $p \geq 1$, the sum over k is bounded, so the first sum in (4.25) is bounded by a constant divided by n^{2p+2}. Similarly, for the second sum in (4.25), we obtain the bound

$$\begin{aligned}
&\frac{C^2}{n^{2p+2}} \sum_{k=-\frac{n}{2}}^{-1} \frac{1}{k^4} \left(\sum_{j\geq 1} \frac{1}{|\frac{k}{n}+j|^{p+1}} + \frac{1}{|\frac{k}{n}-j|^{p+1}} \right)^2 \\
&\leq \frac{C^2}{n^{2p+2}} \sum_{k=-\frac{n}{2}}^{-1} \frac{1}{k^4} \left(\sum_{j\geq 1} \frac{1}{|j-\frac{1}{2}|^{p+1}} + \frac{1}{j^{p+1}} \right)^2 ,
\end{aligned}$$

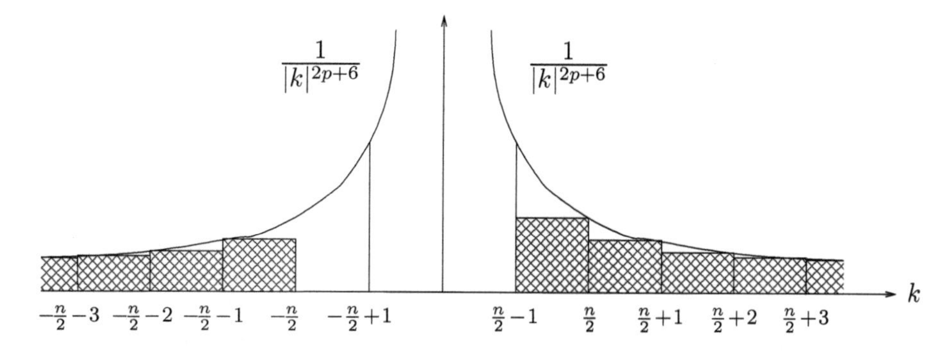

Figure 4.4. *Estimation of a sum by an integral.*

and this term is also bounded by a constant divided by n^{2p+2}. We therefore obtain for the first sum in (4.24) the bound

$$\sum_{k \in K \setminus 0} \frac{1}{k^4} |f_k - \hat{f}_k|^2 \leq \frac{C_1}{n^{2p+2}} \tag{4.26}$$

for some constant C_1. Now for the second sum in (4.24), we use Theorem 4.2 to estimate $|f_k|$ by $|k|^{-(p+1)}$. Then using an integral to estimate the sum as in Figure 4.4, we obtain

$$\begin{aligned}
\sum_{k \notin K} \frac{1}{k^4} |f_k|^2 \leq \sum_{k \notin K} \frac{C^2}{|k|^{2p+6}} &\leq 2C^2 \int_{\frac{n}{2}-1}^{\infty} \frac{1}{k^{2p+6}} dk \\
= \frac{2C^2}{2p+5} \frac{1}{(\frac{n}{2}-1)^{2p+5}} &= \frac{2C^2}{2p+5} \frac{1}{(\frac{1}{2}-\frac{1}{n})^{2p+5}} \frac{1}{n^{2p+5}} \\
\leq \frac{2C^2}{(2p+5)(\frac{1}{4})^{2p+5}} \frac{1}{n^{2p+5}},
\end{aligned}$$

where we used in the last steps that $n \geq 4$. We therefore obtain the estimate

$$\sum_{k \notin K} \frac{1}{k^4} |f_k|^2 \leq \frac{C_2}{n^{2p+5}} \tag{4.27}$$

for some constant C_2. Together with (4.26), we thus obtain

$$\|\mathbf{u} - \mathbf{u}_h\|_2^2 \leq \frac{2\pi C_1}{n^{2p+2}} + \frac{2\pi C_2}{n^{2p+5}}, \tag{4.28}$$

which implies (4.23) for some constant C. □

We see that even though the continuous solution u has two more derivatives than the right-hand-side f because of the integration, the discrete Fourier spectral method only converges as if u had the same regularity as f. The extra regularity is lost because of the quadrature on f, with the first term in (4.24) dominating the overall estimate (4.28). Without the quadrature effects, only the second term in (4.28) would be present, and the method would converge at a higher rate corresponding to the higher regularity of the continuous solution u rather than to the less regular right-hand-side f.

To compare this error estimate with the ones we have obtained for the finite difference and finite volume methods, we use again the grid size $h := \frac{2\pi}{n}$ and obtain from (4.23)

$$||\mathbf{u} - \mathbf{u}_h||_2 = O(h^{p+1}),$$

which is spectral accuracy. It is thus just the regularity of the right-hand-side f that determines the rate of convergence of the spectral method. If f is infinitely differentiable (globally as a periodic function), then the method converges faster than $\mathcal{O}(h^p)$ for every p; this is called *exponential convergence* or *spectral convergence*.

We show in Figure 4.5 the convergence we observe when the periodic one-dimensional Poisson problem (4.2) is solved with the right-hand sides shown in Figure 4.3. We see that the convergence rate is as predicted by Theorem 4.7, except for the third and fifth cases, where the order of convergence is even one higher than expected. These results were produced with the following MATLAB script:

```
p=10;
for j=1:6
  nmax=2^(p+3); [ur,x0]=FFPoisson1d(f{j},0,2*pi,nmax);    % reference solution
  n=2*2.^(1:p);
  for i=1:p,
    [u,x]=FDPoisson1d(f{j},0,2*pi,n(i));                  % finite difference
    errfd(i)=1/sqrt(n(i))*norm(ur(1:nmax/n(i):nmax+1)-u); % solutions
  end;
  for i=1:p;
    [u,x]=FFPoisson1d(f{j},0,2*pi,n(i));                  % discrete spectral
    errff(i)=1/sqrt(nd(i))*norm(ur(1:nmax/n(i):nmax+1)-u);% solutions
  end;
  if j<6
    loglog(nd,errfd,'--',n,errff,'-.',n,1000./n.^j,'-');
    legend('Finite Difference Method','Spectral Method',['h^' num2str(j)])
  else
    loglog(nd,errfd,'-.',n,errff,'-');
    legend('Finite Difference Method','Spectral Method')
  end
  xlabel('number of points'); ylabel('error');
end
```

Here, the functions `f{j}` are defined using the MATLAB script on page 91, and the finite difference solver `FDPoisson1d` and spectral solver `FFPoisson` are implemented as follows:

```
function [u,x]=FDPoisson1d(f,a,b,n);
% FDPOISSON1D solves the 1d Poisson equation with periodic boundary conditions
%   u=Poisson1d(f,a,b,n); solves the 1d poisson equation with right
%   hand side function f and periodic boundary conditions using centered finite
%   differences and n+1 mesh points. Note that f is made to have mean zero.

h=(b-a)/n; x=(a:h:b)';                        % finite difference mesh
e=ones(n,1);
A=spdiags([e -2*e e],-1:1,n,n)/h^2;           % finite difference matrix
A(n,1)=1/h^2;
b=feval(f,x(1:n));                            % check mean in the right
bs=sum(b);                                    % hand side and remove if
if abs(bs)>100*eps,                           % necessary
  warning(['right hand side function mean ' num2str(bs) 'removed']);
  b=b-bs/n;
end;
b(1)=0;                                       % replace redundant equation
A(1,1:n)=ones(1,n);                           % by mean zero constraint
```

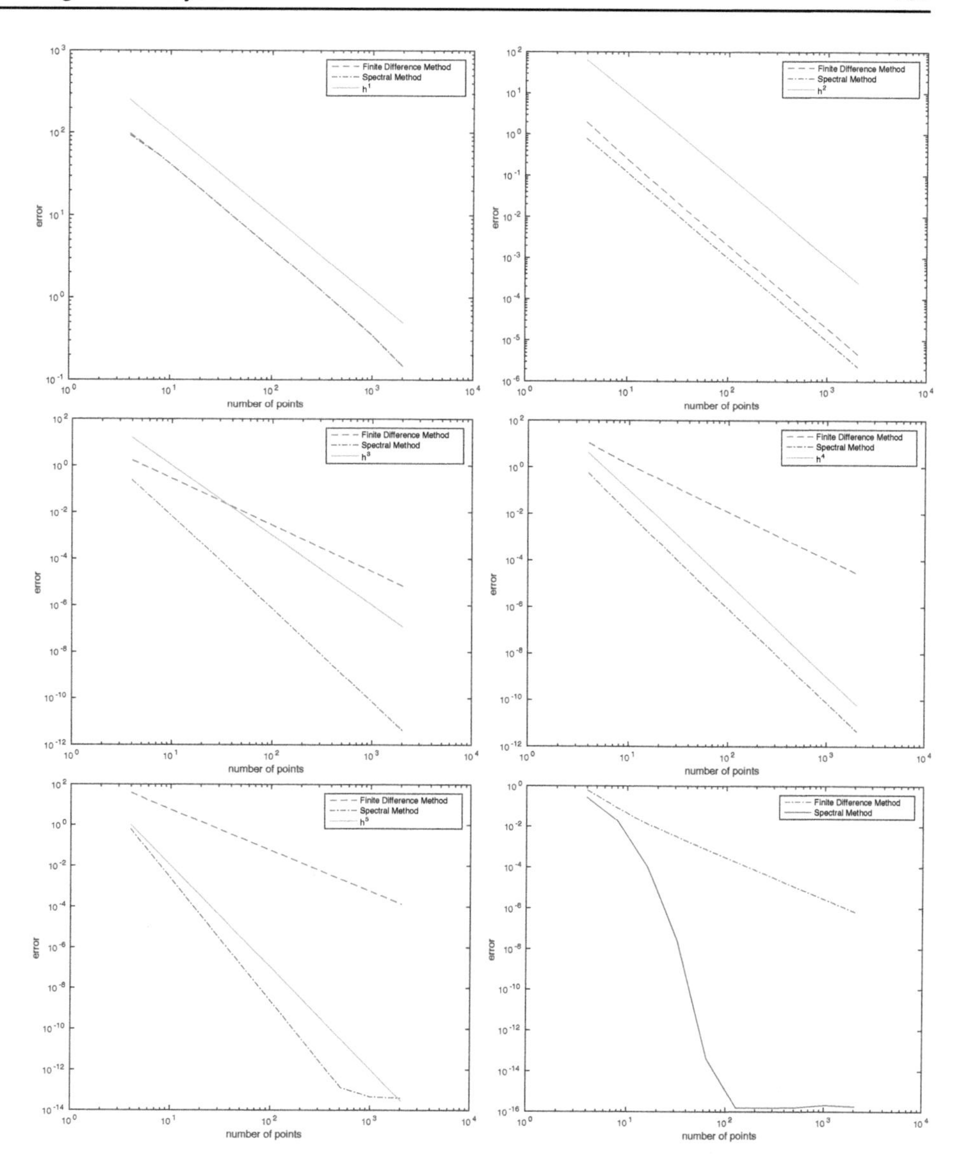

Figure 4.5. *Illustration of the convergence of the discrete spectral Fourier method compared to the finite difference method for right-hand sides with increasing regularity from Figure* 4.3. *One can clearly see that the same spectral code converges faster and faster as the regularity increases.*

```
u=A\b;
u(n+1)=u(1);                                  % solution is periodic
```

```
function [u,x]=FFPoisson1d(f,a,b,n);
% FFPOISSON1D solves the 1d Poisson equation
%   u=FFPoisson1d(f,a,b,n); solves the 1d poisson equation with right
%   hand side function f, boundary values a and b using the fast Fourier
%   transform. Note that f needs to be periodic.
```

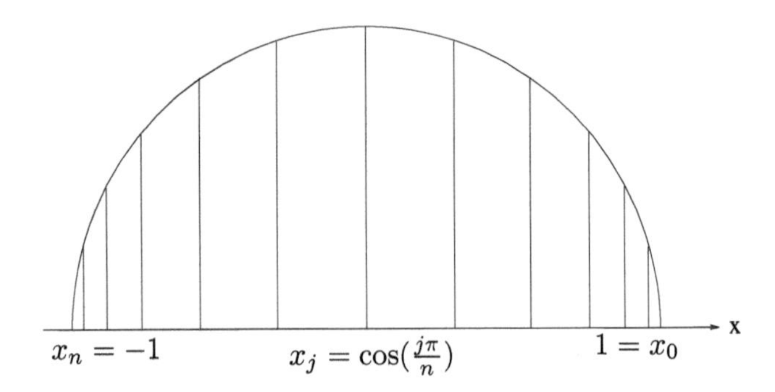

Figure 4.6. *Chebyshev points are the projection of equally distributed points from a circle onto the x-axis.*

```
h=(b-a)/n; x=(a:h:b)';                         % finite difference mesh
b=feval(f,x(1:n));
bh=fft(b);                                     % check mean in rhs and
if abs(bh(1))>100*eps,                         % remove if necessary
  warning(['right hand side function mean ' num2str(bh(1)) 'removed']);
  bh(1)=0;
end;
bhint(2:n/2)=-bh(2:n/2)./([1:n/2-1]'.^2);      % ordering of -k^2 in
bhint(n/2+1:n)=-bh(n/2+1:n)./([-n/2:-1]'.^2); % the FFT implementation
bhint(1)=0;                                    % fix average to 0
u=ifft(bhint);
u(n+1)=u(1);                                   % solution is periodic
u=real(u)';                                    % make numerically real
```

Note that `FFPoisson1d` uses the highly optimized MATLAB built-in functions `fft` and `ifft` as subroutines.

4.4 ▪ Spectral Method Based on Chebyshev Polynomials

For nonperiodic problems, we need to use nonperiodic basis functions for the spectral method, and a typical choice is to use polynomials. Since interpolating polynomials tend to suffer from excessive oscillations when one uses equidistant grid points, a phenomenon that was already pointed out by Runge, it is necessary to redistribute the grid points.

Definition 4.8 (Chebyshev points). *The* Chebyshev points on $[-1, 1]$ *are defined by*

$$x_j = \cos\left(\frac{j\pi}{n}\right), \qquad j = 0, 1, \ldots, n, \qquad \textit{for } x \in [-1, 1]. \tag{4.29}$$

The change of coordinates $x = \cos(\theta)$, which concentrates grid points near the boundary (see Figure 4.6), has a second remarkable property: It changes the periodic

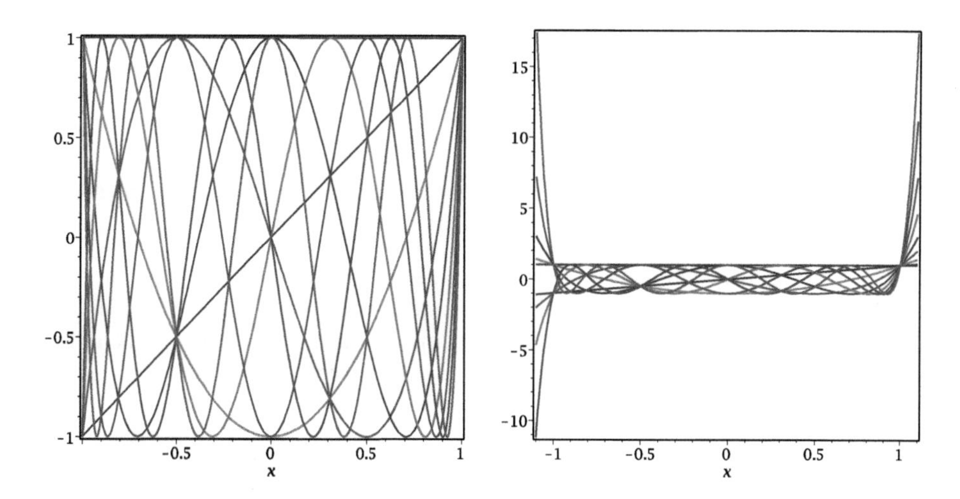

Figure 4.7. *Left: Chebyshev polynomials oscillate on the interval* $[-1, 1]$ *similar to trigonometric functions. Right: The same Chebyshev polynomials on the slightly larger interval* $[-1.1, 1.1]$*, which shows how rapidly they grow outside the interval* $[-1, 1]$.

functions $1, \cos(\theta), \cos(2\theta), \ldots$ into polynomials:

$$\begin{aligned}
T_0 &= 1, \\
T_1 &= \cos(\arccos x) = x, \\
T_2 &= \cos(2\arccos x) = 2\cos^2(\arccos x) - 1 = 2x^2 - 1, \\
&\vdots \\
T_n &= \cos(n \arccos x).
\end{aligned}$$

The polynomials T_k are the Chebyshev polynomials. As one can expect from the change of coordinates, they oscillate like cosines, but they are not periodic. Maple knows the Chebyshev polynomials; with the commands

```
n:=8;
for i from 0 to n do
  p[i]:=simplify(ChebyshevT(i,x));
od;
plot([seq(p[j],j=0..n)],x=-1..1);
```

we obtain the graph of the first few Chebyshev polynomials shown in Figure 4.7. The idea of the Chebyshev spectral method is to look for a solution of the form

$$u(x) = \sum_{k=0}^{\infty} u_k T_k(x). \tag{4.30}$$

In order to obtain an efficient method, we need

1. a fast Chebyshev transform (like the FFT);

2. an easy-to-invert formula for the derivative (like the diagonal differentiation matrix (4.19)).

For the first point, we note that the Chebyshev grid points are $x_j = \cos(\frac{j\pi}{n})$, and hence if we truncate the expansion (4.30) to obtain a fully discrete Chebyshev spectral method, we get

$$u(x_j) = \sum_{k=0}^{n} \hat{u}_k T_k(x_j) = \sum_{k=0}^{n} \hat{u}_k \cos(k \arccos(x_j)) = \sum_{k=0}^{n} \hat{u}_k \cos\left(\frac{kj\pi}{n}\right).$$

This last sum is just a discrete cosine series, and the discrete cosine transform can be computed efficiently with an algorithm similar to the FFT; see Problem 4.3. In addition, we can also expect the truncated sums to converge spectrally to the infinite Chebyshev series because of this relation to the Fourier series.

For the second point, we compute the derivative,

$$\frac{d}{dx}u(x) = \frac{d}{dx}\sum_{k=0}^{n} \hat{u}_n T_k(x) = \sum_{k=0}^{n} \hat{u}_k \frac{d}{dx}T_k(x) \stackrel{!}{=} \sum_{k=0}^{n} \hat{p}_k T_k(x), \tag{4.31}$$

and we thus need a fast method to determine the coefficients $\hat{p}_k$ from the $\hat{u}_k$.

Lemma 4.9. *The coefficients $\hat{p}_k$ and $\hat{u}_k$ are related by the* differentiation matrix $\hat{D}_C$,

$$\begin{pmatrix} \hat{p}_0 \\ \hat{p}_1 \\ \vdots \\ \hat{p}_n \end{pmatrix} = \hat{D}_C \begin{pmatrix} \hat{u}_0 \\ \hat{u}_1 \\ \vdots \\ \hat{u}_n \end{pmatrix}, \tag{4.32}$$

where

$$\hat{D}_C(i,j) = \begin{cases} j-1, & i=1,\ j>1,\ j \text{ even}, \\ 2(j-1), & i>1,\ j>1,\ i+j \text{ odd}. \end{cases}$$

Before proving this lemma, we show for illustration purposes the differentiation matrix for the $n = 5$ case,

$$\hat{D}_C = \begin{bmatrix} 0 & 1 & 0 & 3 & 0 & 5 \\ & 0 & 4 & 0 & 8 & 0 \\ & & 0 & 6 & 0 & 10 \\ & & & 0 & 8 & 0 \\ & & & & 0 & 10 \\ & & & & & 0 \end{bmatrix}.$$

Note that the matrix is singular, as was the case for the discrete Fourier spectral method, but this will be fixed by the boundary conditions; see (4.34). Once this is done, the fact that the matrix is upper triangular means the associated linear systems can be solved easily without using Gaussian elimination, just like the diagonal matrices (4.19) in the Fourier case.

Proof. From $T_k = \cos(k \arccos x)$, we obtain for the derivative

$$T_k'(x) = -\sin(k \arccos x)\left(-\frac{k}{\sin(\arccos x)}\right).$$

We can therefore relate the derivatives of the Chebyshev polynomials to the Chebyshev polynomials themselves for $k > 1$,

$$\begin{aligned}\frac{1}{k+1}T'_{k+1}(x) - \frac{1}{k-1}T'_{k-1}(x) &= \frac{\sin((k+1)\arccos x) - \sin((k-1)\arccos x)}{\sin(\arccos(x))} \\ &= 2\cos(k\arccos x) = 2T_k(x), \end{aligned} \tag{4.33}$$

where we used the fact that $\sin(A+B) - \sin(A-B) = 2\cos(A)\sin(B)$. Using the explicit expressions $T_0 = 1$, $T_1 = x$, and $T_2 = 2x^2 - 1$ for the first three Chebyshev polynomials, we can write the relation (4.33) in matrix form,

$$(T_0, T_1, \ldots, T_{n-1}) = (T'_1, T'_2, \ldots, T'_n) \underbrace{\begin{bmatrix} 1 & 0 & -\frac{1}{2} & & & \\ & \frac{1}{4} & 0 & -\frac{1}{4} & & \\ & & \frac{1}{6} & 0 & -\frac{1}{6} & \\ & & & \ddots & \ddots & \ddots \end{bmatrix}}_{=:M}.$$

We can thus rewrite the relation (4.31),

$$\begin{aligned}(T'_1, T'_2, \ldots, T'_n)\begin{pmatrix}\hat{u}_1\\ \hat{u}_2\\ \vdots\\ \hat{u}_n\end{pmatrix} &\stackrel{!}{=} (T_0, T_1, \ldots, T_{n-1})\begin{pmatrix}\hat{p}_0\\ \hat{p}_1\\ \vdots\\ \hat{p}_{n-1}\end{pmatrix} \\ &= (T'_1, T'_2, \ldots, T'_n)M\begin{pmatrix}\hat{p}_0\\ \hat{p}_1\\ \vdots\\ \hat{p}_{n-1}\end{pmatrix}.\end{aligned}$$

Comparing the coefficients, we thus obtain

$$\begin{bmatrix} 0 & & \\ \vdots & M^{-1} & \\ 0 & \ldots & 0 \end{bmatrix}\begin{pmatrix}\hat{u}_0\\ \hat{u}_1\\ \vdots\\ \hat{u}_n\end{pmatrix} = \begin{pmatrix}\hat{p}_0\\ \hat{p}_1\\ \vdots\\ \hat{p}_n\end{pmatrix}.$$

The inverse of M is easy to compute using the factorization

$$\begin{aligned} M &= \begin{bmatrix}\frac{1}{2} & & & \\ & \frac{1}{4} & & \\ & & \frac{1}{6} & \\ & & & \ddots\end{bmatrix}(I - N)\begin{bmatrix}2 & & & \\ & 1 & & \\ & & \ddots & \\ & & & 1\end{bmatrix} \\ &= \begin{bmatrix}\frac{1}{2} & & & \\ & \frac{1}{4} & & \\ & & \frac{1}{6} & \\ & & & \ddots\end{bmatrix}\begin{bmatrix}2 & 0 & -1 & & & \\ & 1 & 0 & -1 & & \\ & & 1 & 0 & -1 & \\ & & & \ddots & \ddots & \ddots\end{bmatrix}, \end{aligned}$$

where the matrix N is given by

$$N := \begin{bmatrix}0 & 0 & 1 & & \\ & 0 & 0 & 1 & \\ & & \ddots & \ddots & \ddots\end{bmatrix}.$$

We can now compute M^{-1}, using the fact that $(I-N)^{-1} = I + N + N^2 + \cdots$,

$$M^{-1} = \begin{bmatrix} \frac{1}{2} & & & \\ & 1 & & \\ & & \ddots & \\ & & & 1 \end{bmatrix} (I-N)^{-1} \begin{bmatrix} 2 & & & \\ & 4 & & \\ & & 6 & \\ & & & \ddots \end{bmatrix}$$

$$= \begin{bmatrix} \frac{1}{2} & & & \\ & 1 & & \\ & & \ddots & \\ & & & 1 \end{bmatrix} \begin{bmatrix} 1 & 0 & 1 & 0 & 1 & \cdots & \\ & 1 & 0 & 1 & 0 & 1 & \cdots \\ & & \ddots & \ddots & \ddots & \ddots & \ddots \end{bmatrix} \begin{bmatrix} 2 & & & \\ & 4 & & \\ & & 6 & \\ & & & \ddots \end{bmatrix}$$

$$= \begin{bmatrix} \frac{1}{2} & & & \\ & 1 & & \\ & & \ddots & \\ & & & 1 \end{bmatrix} \begin{bmatrix} 2 & 0 & 6 & 0 & 10 & \cdots \\ & 4 & 0 & 8 & 0 & \cdots \\ & & 6 & 0 & 10 & \cdots \\ & & & \ddots & \ddots & \ddots \end{bmatrix} = \begin{bmatrix} 1 & 0 & 3 & 0 & 5 & \cdots \\ & 4 & 0 & 8 & 0 & \cdots \\ & & 6 & 0 & 10 & \cdots \\ & & & \ddots & \ddots & \ddots \end{bmatrix},$$

which completes the proof of the lemma. □

The following MATLAB code implements the Chebyshev spectral method for solving $u_{xx} = f$ on the interval $[-1,1]$. Here, instead of implementing the discrete cosine transform from scratch, we take the values of f at Chebyshev points and do an even extension before feeding it to the built-in MATLAB function `fft`; see Problem 4.4. The inverse cosine transform is computed similarly. Next, the boundary conditions are given by the two additional equations

$$\begin{aligned} u_L = u(-1) &= \sum_{k=0}^{n} \hat{u}_k T_k(-1) = \hat{u}_0 - \hat{u}_1 + - \cdots + (-1)^n \hat{u}_n, \\ u_R = u(1) &= \sum_{k=0}^{n} \hat{u}_k T_k(1) = \hat{u}_0 + \hat{u}_1 + \cdots + \hat{u}_n. \end{aligned} \tag{4.34}$$

These equations will replace the last two rows of $\hat{D}_C^2$, which are identically zero.

```
function [x,u]=ChebyshevPoisson1d(f,uL,uR,n)
% CHEBYSHEVPOISSON1D Chebyshev spectral solver for the 1d Poisson equation
%   [x,u]=ChebyshevPoisson1d(f,uL,uR,n) solves uxx=f on [-1,1] with
%   u(-1)=uL, u(1)=uR using n points (excluding end points)

n=n+1;                                        % Number of intervals
xi=(0:pi/n:pi)';                              % Interpolation points
ff=f(cos(xi));                                % Sample f at Chebyshev pts
fhat=fft([ff(1:end-1);ff(end:-1:2)])/(2*n);   % Discrete Cosine Transform
fhat=fhat(1:n-1);                             % (first n-1 coeffs only)
fhat(2:end)=fhat(2:end)*2;

D=zeros(n+1);                                 % Differentiation matrix
for j=1:n,
  D(j:-2:1,j+1)=2*j;
end;
D(1,:)=D(1,:)/2;
A=D^2;                                        % Square to get Laplacian
```

```
A(end-1,:)=(-1).^(0:n);                      % Left boundary condition
A(end,:)=ones(1,n+1);                        % Right boundary condition
b=[fhat; uL; uR];                            % Right hand side
uhat=A\b;
x=cos((0:pi/n:pi)');                         % Sample points
uuhat=0.5*[uhat(1:end-1);uhat(end:-1:2)];    % Inverse DCT
uuhat(1)=uuhat(1)*2; uuhat(n+1)=uuhat(n+1)*2;
u=ifft(uuhat)*2*n;
u=u(1:n+1);
```

To illustrate the Chebyshev spectral method, we use it to solve the problem $u_{xx} = f$ on $[-1, 1]$, where f, u_L, and u_R are chosen so that we have the exact solution

$$u(x) = \frac{1}{1 + 25x^2}.$$

In Figure 4.8, we plot the numerical solution obtained by the Chebyshev spectral and finite difference methods for the same number N of degrees of freedom for $N = 4, 8, 16, 32, 64$. For very small N, both methods perform poorly; the finite difference method only becomes reasonably accurate for $N \geq 16$ and the Chebyshev method for $N \geq 32$. To understand this, consider the plots in Figure 4.9, where we show the function f and its discrete cosine transform. We see that f drops off sharply near $x = 0$; if the finite difference method does not take enough samples of f near this trough, i.e., when N is not large enough, then it cannot possibly get an accurate answer, just as we observed in the solution plots. Similarly, the largest entries in the discrete cosine transform of f occur for $k \leq 30$, which is the minimum number of degrees of freedom needed to obtain a reasonably accurate solution. Beyond about $k = 30$, we see that the Chebyshev spectral method converges exponentially thanks to the regularity of the right-hand-side f, and it rapidly converges to within machine precision; see the bottom right panel of Figure 4.8.

4.5 ▪ Concluding Remarks

Spectral methods are based on a global expansion of the solution using a set of basis functions which have support throughout the domain. We have barely scratched the surface in our treatment: For more details and techniques, we refer the reader to the excellent volume by Trefethen [61]. In the form we have seen them in this chapter, spectral methods have the following advantages and disadvantages, denoted by plus and minus signs, respectively:

+ For smooth solutions, the spectral method gives highly accurate approximations, much more accurate than the finite difference or finite volume methods can.

– The spectral method in the form presented here can only be used on simple, rectangular geometries.

We will see, however, in the next section on the finite element method how arbitrary geometries can be decomposed into regular patches for the numerical approximation of PDEs, and this approach can be combined with the spectral method on each patch, leading to the spectral finite element method.

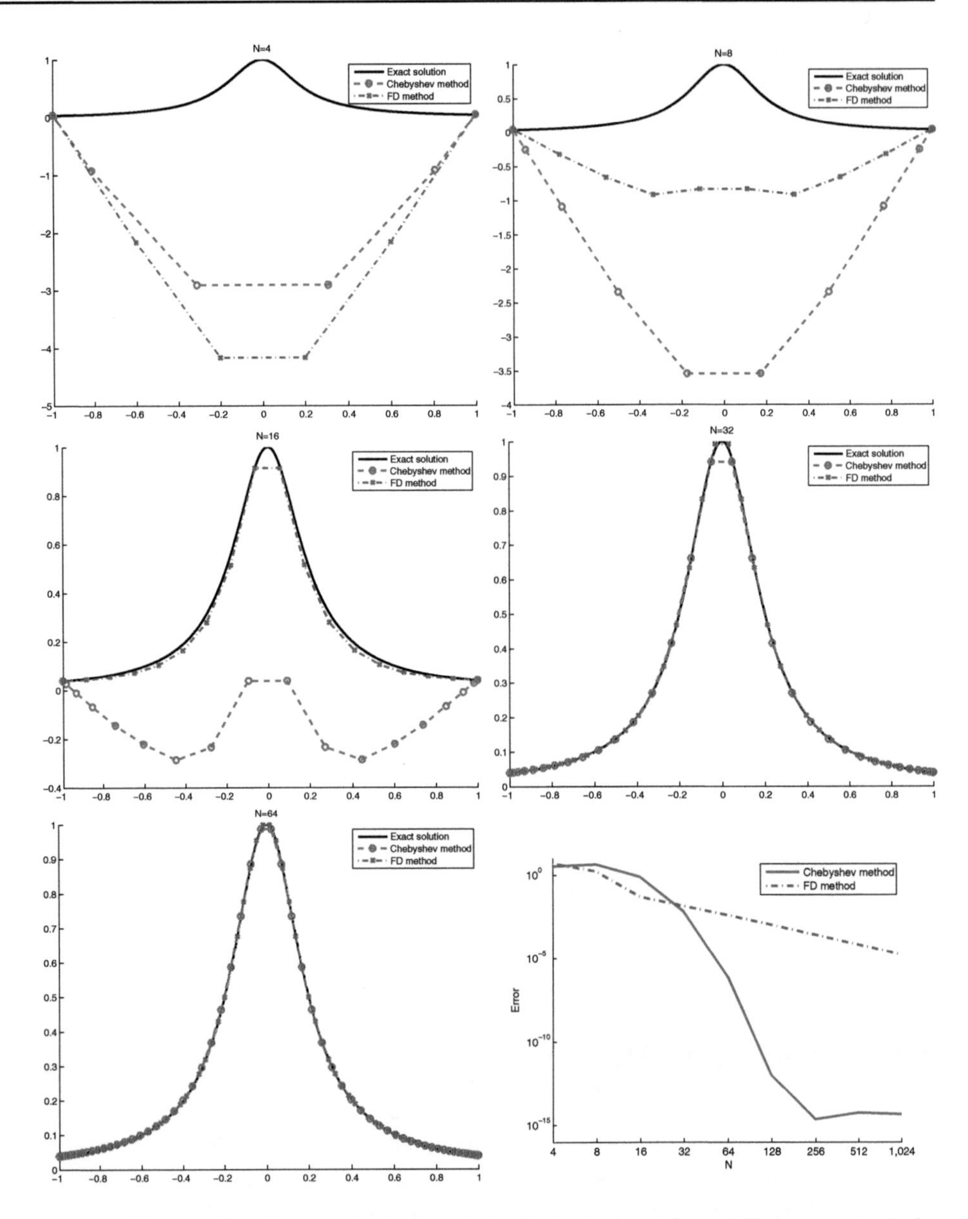

Figure 4.8. *Numerical solution of the Chebyshev and finite difference methods for* $N = 4, 8, 16, 32, 64$ *degrees of freedom. Bottom right: error of the two methods for different* N.

4.6 ▪ Problems

Problem 4.1 (periodic Poisson problem). We search $u : [0, 2\pi] \to \mathbb{R}$ solution of

$$\begin{aligned} u_{xx} &= f(x), \\ u(0) &= u(2\pi), \\ u_x(0) &= u_x(2\pi). \end{aligned}$$

1. Find the conditions on f such that this equation has a solution. Is the solution then unique? (Hint: Use Fourier series.)

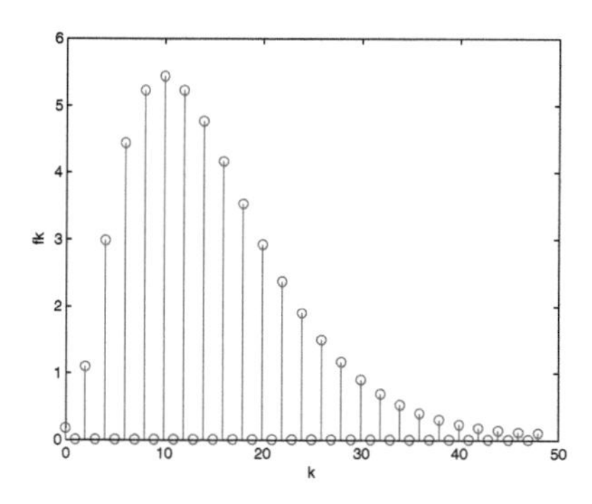

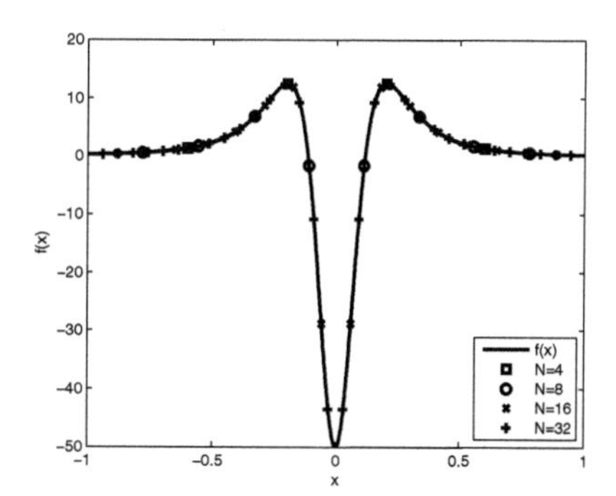

Figure 4.9. *Left: The magnitude of the first* 50 *discrete cosine coefficients. Right: The right-hand-side function f and the points sampled by the finite difference method for different values of N.*

2. Discretize the problem using a finite difference method and check if similar conditions hold for the existence and uniqueness of the discrete solution.

3. Implement the method in MATLAB using the header

```
function [u,x]=FDPoisson1d(f,a,b,n);
% FDPOISSON1D solves the 1d Poisson equation with periodic boundary conditions
%   u=Poisson1d(f,a,b,n); solves the 1d poisson equation with
%   right hand side function f and periodic boundary conditions at a and b
%   using centered finite differences and n+1 mesh points. Note that f is
%   made to have mean zero.
```

4. Implement the discrete Fourier spectral method in MATLAB using the built-in `fft` function. Use the header

```
function [u,x]=FFPoisson1d(f,a,b,n);
% FFPOISSON1D solves the 1d Poisson equation.
%   u=FFPoisson1d(f,a,b,n); solves the 1d poisson equation with right
%   hand side function f, periodic boundary values at a and b using the
%   fast Fourier transform. Note that f is made to have mean zero.
```

5. Compare the performance of the two methods for

$$f(x) = \frac{1}{2 - \cos(x)}$$

modified such that a solution exists. Use $n = 4, 8, 16, 32, 64$ points, and plot graphs of the error using a logarithmic scale. Can you explain the kind of convergence you observe?

Problem 4.2 (Chebyshev polynomials). For $n = 0, 1, 2, \ldots$ and $x \in [-1, 1]$, the Chebyshev polynomials are defined by

$$T_n(x) = \cos(n \arccos(x)).$$

Prove the following properties of the Chebyshev polynomials:

1. The functions $T_n(x)$ satisfy the recurrence relation

$$T_0(x) = 1, \quad T_1(x) = x, \quad T_{n+1} = 2xT_n(x) - T_{n-1}(x).$$

   Hence, T_n is a polynomial of degree n whose leading coefficient is 2^{n-1}.

2. $|T_n(x)| \leq 1$ for $x \in [-1, 1]$.

3. $T_n \cos(\frac{k\pi}{n})) = (-1)^k$ for $k = 0, 1, \ldots, n$.

4. $T_n(\cos(\frac{(2k+1)\pi}{2n})) = 0$ for $k = 0, 1, \ldots, n-1$.

5. The polynomials $T_n(x)$ are orthogonal for the weight function $1/\sqrt{1-x^2}$, i.e.,

$$\int_{-1}^{1} \frac{1}{\sqrt{1-x^2}} T_n(x) T_m(x) dx = \begin{cases} \pi & \text{if } n = m = 0, \\ \frac{\pi}{2} & \text{if } n = m \neq 0, \\ 0 & \text{if } n \neq m. \end{cases}$$

Problem 4.3 (calculating the discrete cosine transform via the FFT). Recall that the discrete Fourier transform of a vector $(F_j)_{j=0}^{N-1}$ is given by $(\hat{f}_k)_{k=0}^{N-1}$, where

$$\hat{f}_k = \frac{1}{N} \sum_{j=0}^{N-1} F_j \exp(-2\pi ijk/N).$$

With this definition, we have the reconstruction formula

$$F_j = \sum_{k=0}^{N-1} \hat{f}_k \exp(2\pi ijk/N), \qquad j = 0, 1, \ldots, N-1.$$

1. Let $N = 2n$ be even, and assume that the vector $(F_j)_{j=0}^{N-1}$ is real and satisfies $F_{N-j} = F_j$. Show that $\hat{f}_{N-k} = \hat{f}_k$ for $k = 1, \ldots, n$ and that

$$F_j = \hat{f}_0 + (-1)^j \hat{f}_n + 2 \sum_{k=1}^{n-1} \hat{f}_k \cos(jk\pi/n).$$

2. Explain how to compute the discrete cosine transform using FFT as a subroutine.

Problem 4.4 (discrete Chebyshev spectral method). The goal of this question is to implement the discrete Chebyshev spectral method to solve the problem

$$\begin{cases} u_{xx} = f & \text{in } \Omega = (-1, 1), \\ u(-1) = u_L, \\ u'(1) = g_R. \end{cases} \tag{4.35}$$

1. Using an appropriate mesh point distribution, show that a fast Chebyshev transform can be obtained by using a fast cosine transform. Implement the following MATLAB functions for the fast cosine transform and its inverse:

```
function y=FCT(x)
% FCT gives the fast cosine transform using the fft
```

```
function y=IFCT(x)
% IFCT gives the fast inverse cosine transform using the ifft
```

2. Using the property $T_n(1) = 1$ and the parity of the Chebyshev polynomials, obtain an equation for the Dirichlet boundary conditions.

3. Show that $T_n'(1) = n^2$ and obtain an equation for the Neumann boundary condition.

4. Modify the differentiation matrix seen in Lemma 4.9 to take into account the boundary conditions.

5. Implement the discrete Chebyshev spectral method to solve (4.35) and test it for different values of f, u_L, and g_R.

```
function [u,x]=ChebyshevPoisson1dNeumann(f,ul,gr,n)
% CHEBYSHEVPOISSON1DNEUMANN solves the Poisson equation
%   [u,x]=ChebyshevPoisson1dNeumann(f,ul,gr,n) solves the one
%   dimensional Poisson equation on the interval (-1,1) with
%   right hand side function f, Dirichlet boundary condition
%   ul on the left and Neumann boundary gr on the right, using
%   n grid points. The solution is returned in u with associated
%   grid points x.
```

Chapter 5

The Finite Element Method

Die Endpunkte dieser Ordinaten bilden die Ecken von ebenen Dreiecken, aus welchen die kleinste Fläche bestehen soll und von deren Projection auf die Ebene der $\xi\eta$ die eine Hälfte dieser Dreiecke in der Figur schraffirt sich zeigt.[a]

Karl Schellbach, Probleme der Variationsrechnung, 1852

Das wesentliche der neuen Methode besteht darin, dass nicht von den Differentialgleichungen und Randbedingungen des Problems, sondern direkt vom Prinzip der kleinsten Wirkung *ausgegangen wird, aus welchem ja durch Variation jene Gleichungen und Bedingungen gewonnen werden können.*[b]

Walther Ritz, Theorie der Transversalschwingungen einer quadratischen Platte mit freien Rändern, 1909

Из приближенных методов решений широкое применение получил в последнее время метод Ритца [6, 7]. Этот метод сводится вкратце к следующему.[c]

Boris Grigoryevich Galerkin, Rods and Plates. Series occurring in various questions concerning the elastic equilibrium of rods and plates, 1915

Un intérêt immédiat de la formulation variationnelle est qu'elle a un sens si la solution u est seulement une fonction de $C^1(\bar{\Omega})$, contrairement à la formulation "classique" qui requiert que u appartienne à $C^2(\bar{\Omega})$. On pressent donc déjà qu'il est plus simple de résoudre [la formulation variationnelle] que [la formulation classique], puisqu'on est moins exigeant sur la régularité de la solution.[d]

Grégoire Allaire, Analyse numérique et optimisation, Éditions de l'École Polytechnique, 2005

---

[a]The end points of the ordinates form corners of planar triangles, which represent the smallest areas, and the projection on the plane of the $\xi\eta$ of half of these triangles is shown shaded; see Figure 5.1 (left).

[b]The most important feature of the new method is that one does not start with the differential equations and boundary conditions of the problem but rather the principle of least action, from which by variation the differential equations and boundary conditions can be obtained.

[c]Among approximate solution methods, the method of Ritz has gained widespread application in recent times. Briefly, this method consists of the following.

[d]An immediate advantage of the variational formulation is that it makes sense if the solution u is only a function in $C^1(\bar{\Omega})$, in contrast to the classical formulation, which requires that u be in $C^2(\bar{\Omega})$. One can thus already foresee that it is easier to solve the variational form compared to the classical one since it is less demanding on the regularity of the solution.

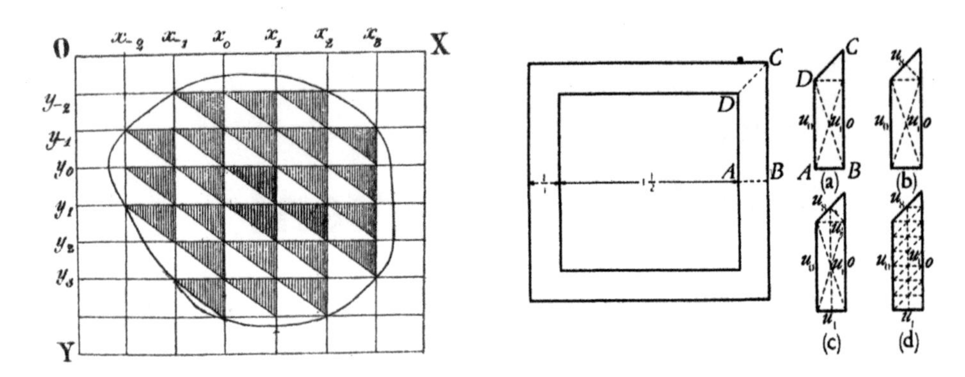

Figure 5.1. *Left: the two-dimensional triangular finite element space in Schellbach* 1852. *Right: finite element computation by Courant which he presented on May* 3, 1941 *in front of the AMS during a meeting in Washington; see* [8] *for a published transcript. Reprinted with permission from the American Mathematical Society.*

The finite element method is the most flexible method for the numerical approximation of PDEs, and it is also the one with the most solid mathematical foundation since it is directly based on the techniques of the calculus of variations. The essential ideas go back to Walther Ritz [49], who was the first in 1908 to introduce the systematic use of finite dimensional approximations of the function spaces in which the solution lives. Ritz focused on the minimization formulation; see the quote above from 1909, where the approximation idea is very natural: Instead of minimizing over all functions from the entire infinite dimensional space, one just minimizes over a finite dimensional subspace. The approximate solution does not generally coincide with the global minimum, but the approximation becomes better and better as the dimension of the approximation space increases, provided that appropriate functions are chosen; see [49] and [50] for eigenvalue problems, where Ritz uses his physical intuition to choose known eigenfunctions of one-dimensional vibration problems to approximate two-dimensional vibration problems. Galerkin then proposed to use the the same finite dimensional approach as Ritz, but instead of going to the minimization formulation, he works directly on the differential equation [20], using an approximation space spanned by sine and cosine functions. Prior to the work of Galerkin, affine functions on a triangular mesh had already been proposed by Schellbach in 1852 to solve a specific minimization problem in the context of minimal surface computations. By refining the mesh and passing to the limit, he obtained the corresponding differential equation; see the quote above and the left panel of Figure 5.1. In 1922, Hurwitz and Courant also used affine functions on a triangular mesh in a footnote in [31] as a means to prove existence of a solution, but this auxiliary construction disappeared already in the second edition. Courant, however, then revived this approach as a fundamental solution procedure in a presentation in front of the AMS in 1941, emphasizing that the main achievement was really due to Ritz and giving a concrete practical example of how the method works on triangular meshes [8]; see also Figure 5.1 on the right.[19] The name of the method appears for the first time in the title of a proceedings paper by Clough [7], who reinvented the method with collaborators [62], without being aware of the earlier work in the mathematical community. For more information about the historical development of the finite element method, see [22] and the references therein.

[19]The square-shaped domain actually represents a cross section of a submarine, and by symmetry, only $1/8$ of the domain needs to be computed (ABCD).

5.1 ▪ Strong Form, Weak or Variational Form, and Minimization

The finite element method is based on the calculus of variations, which is best explained with a simple example. Consider the Poisson equation in one spatial dimension,

$$\begin{cases} -u'' = f & \text{in } \Omega = (0,1), \\ u(0) = 0, \\ u(1) = 0, \end{cases} \tag{5.1}$$

where f is a continuous and bounded function, and we thus search for a twice continuously differentiable solution u. This formulation of the problem is called the *strong form* of the problem.

In order to obtain a finite element discretization of (5.1), one does not work directly with the strong form but rather a reformulation of the problem shown below.[20] One introduces a function space V, where we choose here[21]

$$V := \{v : v \text{ continuous}, v' \text{ piecewise continuous and bounded in } \Omega,\ v(0) = v(1) = 0\}. \tag{5.2}$$

If u is a solution of the strong form (5.1), then multiplying (5.1) by any function $v \in V$ and integrating over the domain Ω, we obtain from integration by parts

$$-\int_0^1 u''(x)v(x)dx = -u'(x)v(x)\Big|_0^1 + \int_0^1 u'(x)v'(x)dx = \int_0^1 f(x)v(x)dx.$$

If we define the inner product

$$(u,v) := \int_0^1 u(x)v(x)dx, \tag{5.3}$$

we obtain the so-called *weak form* or *variational form* of the original problem (5.1).

Definition 5.1 (weak or variational form).

$$\textit{Find } u \in V \textit{ such that } (u', v') = (f, v),\ \forall v \in V. \tag{5.4}$$

We have already seen the idea of multiplying the equation by a function followed by integration for spectral methods (see (4.6)), and Galerkin did indeed motivate this approach first using a sine series in 1915; see [20, pp. 169–171].

Ritz started, however, in 1908 with a minimization problem [49]; see also the quote at the beginning of the chapter, which is very natural from a physical point of view and was also the starting point for the calculus of variations. The important physical functional for our simple example (5.1) is

$$F(v) := \frac{1}{2}(v', v') - (f, v), \quad \forall v \in V. \tag{5.5}$$

---

[20]Historically, in the calculus of variations by Euler and Lagrange, the strong form was obtained at the end, so we are going backward here.

[21]We follow here at the beginning the simplified approach of Grégoire Allaire with this choice of V. To use the powerful tools of analysis for variational formulations, V should be a Hilbert space, and we will later use the more suitable choice $V = H_0^1(\Omega)$, which is defined in section 5.4.

To see why, let u be a solution of the variational form (5.4) of the original problem (5.1). We then have for any function $v \in V$ that

$$\begin{aligned} F(u+v) &= \frac{1}{2}(u'+v', u'+v') - (f, u+v) \\ &= \frac{1}{2}(u', u') - (f, u) + \underbrace{(u', v') - (f, v)}_{=0,u \text{ solution of (5.4)}} + \underbrace{\frac{1}{2}(v', v')}_{\geq 0} \\ &\geq \frac{1}{2}(u', u') - (f, u) = F(u). \end{aligned}$$

Hence, if u is a solution of the variational form (5.4), it also minimizes the functional F. We can therefore reformulate the problem also as a minimization problem.

Definition 5.2 (minimization problem).

$$\textit{Find } u \in V \textit{ such that } F(u) \textit{ is minimal.} \tag{5.6}$$

We have so far shown that if u is a solution of the strong form (5.1) of our problem, then u is also a solution of the weak form (5.4) of our problem, and if u is a solution of the weak form (5.4), then u is also solution of the minimization problem (5.6). We now show that the converse is also true when u has sufficient regularity; this is how the calculus of variations was developed. Suppose that u is a solution of the minimization problem (5.6). This implies that

$$F(u) \leq F(u + \varepsilon v) \quad \forall v \in V \text{ because } u + \varepsilon v \in V. \tag{5.7}$$

We consider now the function $F(u+\varepsilon v)$ as a function[22] of ε and expand the expression,

$$\begin{aligned} F(u+\varepsilon v) &= \frac{1}{2}(u'+\varepsilon v', u'+\varepsilon v') - (f, u+\varepsilon v) \\ &= \frac{1}{2}(u', u') + \varepsilon(u', v') + \frac{\varepsilon^2}{2}(v', v') - (f, u) - \varepsilon(f, v). \end{aligned}$$

Since F has a minimum at u, this function has a minimum at $\varepsilon = 0$, which implies $\frac{\partial}{\partial \varepsilon} F(u + \varepsilon v)\big|_{\varepsilon=0} = 0$. We therefore obtain for all $v \in V$

$$\frac{\partial}{\partial \varepsilon} F(u+\varepsilon v)\bigg|_{\varepsilon=0} = (u', v') - (f, v) = 0 \quad \Longrightarrow \quad (u', v') = (f, v), \tag{5.8}$$

which shows that u is a solution of the weak form (5.4). Finally, let u be a solution of the weak form (5.4), and assume in addition that u has a bit more regularity, e.g., if $u \in \mathcal{C}^2(\Omega) \subset V$. We then obtain from the weak form, using again integration by parts, that

$$(u', v') = (f, v) \qquad \forall v \in V \quad \Longrightarrow \quad -\int_0^1 vu'' + \cancel{u'v\Big|_0^1} = (f, v),$$

or equivalently

$$\int_0^1 (u'' + f) v dx = 0 \qquad \forall v \in V.$$

[22]It took Euler 20 years to introduce this way of thinking to clearly explain the derivatives introduced by Lagrange in the calculus of variations.

We now show by contradiction that this implies the strong form of the problem,

$$u'' + f = 0 \text{ on } (0, 1).$$

Suppose $u'' + f \neq 0$: Then by continuity, there exists $x \in (0, 1)$ such that $u'' + f \neq 0$ in a neighborhood of x. If we then choose a function v that has the same sign in this neighborhood and is zero everywhere else, we obtain $\int_0^1 (u'' + f)v dx > 0$, which is a contradiction. Therefore, we must have that $-u'' = f$ and u is solution of the strong form (5.1) of our problem. By showing these equivalences, we have traveled on the "highway" of the calculus of variations (see also the quote of Ritz at the beginning of this chapter) and obtained the following theorem.

Theorem 5.3 (equivalence of the strong, weak, and minimization forms). *The strong form of the problem* (5.1) *implies the weak or variational form* (5.4)*, which in turn is equivalent to the minimization* (5.6)*. If, in addition, the solution is sufficiently regular with $u \in C^2(\Omega)$, then both the minimization* (5.6) *and the weak or variational forms* (5.4) *imply the strong form* (5.1).

5.2 ▪ Discretization

Following the idea of Ritz, we restrict ourselves to a finite dimensional subspace of V to obtain a numerical method. We introduce the subspace V_h of V defined by

$$V_h := \text{span}\{\varphi_1, \ldots, \varphi_n\} \subset V, \tag{5.9}$$

where φ_j, $j = 1, \ldots, n$ are a set of given functions. If we write the *weak or variational form* using only functions from the approximate subspace V_h, we obtain the Galerkin approximation.

Definition 5.4 (Galerkin approximation).

$$\textit{Find } u_h \in V_h \textit{ such that } (u_h', v_h') = (f, v_h) \quad \forall v_h \in V_h. \tag{5.10}$$

If we write u_h as a linear combination of the functions in V_h,

$$u_h = \sum_{j=1}^{n} u_j \varphi_j,$$

requiring (5.10) be satisfied for all $v_h \in V_h$ is equivalent by linearity to requiring it to be satisfied for all functions φ_i, $i = 1, \ldots, n$. The Galerkin approximation (5.10) then implies

$$\left(\sum_{j=1}^{n} u_j \varphi_j', \varphi_i' \right) = \sum_{j=1}^{n} u_j (\varphi_j', \varphi_i') = (f, \varphi_i),$$

and we recognize that this is just a linear system of equations,

$$K\mathbf{u} = \mathbf{f} \tag{5.11}$$

with the *stiffness matrix*

$$K_{ij} := (\varphi_i', \varphi_j') \tag{5.12}$$

and $f_i := (f, \varphi_i)$.

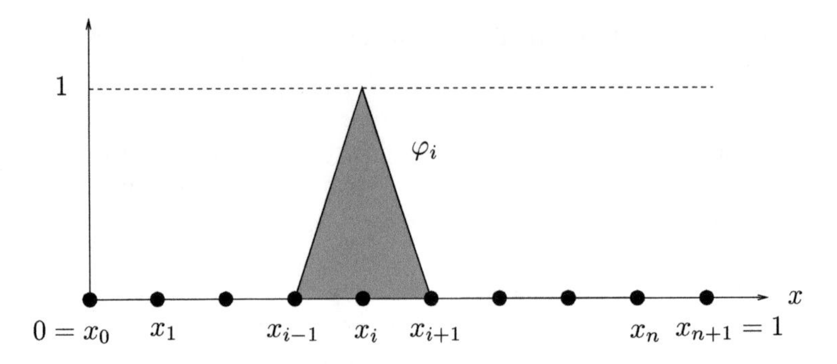

Figure 5.2. *Hat function φ_i around the mesh point x_i.*

Instead of computing the integrals for f_i or an approximation by quadrature, one can also first approximate the function f as a linear combination of the functions φ_j that span V_h,

$$f(x) \approx \tilde{f}(x) := \sum_{j=1}^{n} \tilde{f}_j \varphi_j(x) \in V_h.$$

Using $\tilde{f}$ instead of f in the Galerkin approximation (5.10), we obtain

$$\left(\sum_{j=1}^{n} \tilde{f}_j \varphi_j, \varphi_i \right) = \sum_{j=1}^{n} \tilde{f}_j (\varphi_j, \varphi_i) =: M\tilde{\mathbf{f}},$$

where we see a second matrix appear, the so-called *mass matrix*:

$$M_{ij} = (\varphi_i, \varphi_j). \tag{5.13}$$

In this case, the linear system obtained from the Galerkin approximation would be of the form

$$K\mathbf{u} = M\tilde{\mathbf{f}}. \tag{5.14}$$

Example 5.5. We introduce on the domain $\Omega = (0,1)$ a mesh $0 = x_0 < x_1 < x_2 < \cdots < x_{n+1} = 1$ and let V_h be the space of piecewise linear *hat functions* φ_i (see Figure 5.2) such that

$$\varphi_i' = \begin{cases} \frac{1}{x_i - x_{i-1}} = \frac{1}{h_i}, & x_{i-1} < x < x_i, \\ \frac{-1}{x_{i+1} - x_i} = -\frac{1}{h_{i+1}}, & x_{i-1} < x < x_i, \\ 0 & \text{otherwise.} \end{cases}$$

We can then easily compute the corresponding stiffness matrix K and mass matrix M by evaluating the integrals (5.12) and (5.13),

$$K_{ij} = (\varphi_i', \varphi_j') = \begin{cases} -\frac{1}{h_i}, & j = i-1, \\ \frac{1}{h_i} + \frac{1}{h_{i+1}}, & j = i, \\ -\frac{1}{h_{i+1}}, & j = i+1, \\ 0 & \text{otherwise,} \end{cases}$$

$$M_{ij} = (\varphi_i, \varphi_j) = \begin{cases} \frac{1}{6} h_i, & j = i-1, \\ \frac{1}{3}(h_i + h_{i+1}), & j = i, \\ \frac{1}{6} h_{i+1}, & j = i+1, \\ 0 & \text{otherwise.} \end{cases}$$

If all the mesh cells have the same size, $h_i = h$ for all i, then

$$K = \frac{1}{h}\begin{bmatrix} 2 & -1 & & & \\ -1 & 2 & -1 & & \\ & & \ddots & & \\ & & -1 & 2 & -1 \\ & & & -1 & 2 \end{bmatrix}, \quad M = \frac{h}{6}\begin{bmatrix} 4 & 1 & & & \\ 1 & 4 & 1 & & \\ & & \ddots & & \\ & & 1 & 4 & 1 \\ & & & 1 & 4 \end{bmatrix},$$

and we see that the stiffness matrix K we obtain corresponds to the same approximation finite differences would give us (see (2.5)), up to a factor of h, which we find, however, in the mass matrix. It is thus important to realize that the discrete linear system (5.11) or (5.14) obtained from a Galerkin approximation would need a division by h in one spatial dimension to be interpreted directly as an approximation of the differential equation.

Lemma 5.6. *The stiffness matrix K and the mass matrix M are* symmetric and positive definite.

Proof. We start with the stiffness matrix K. From its definition, the symmetry follows directly, $K_{ij} = (\varphi_i', \varphi_j') = (\varphi_j', \varphi_i') = K_{ji}$. To see that K is also positive definite, we have to show for any vector $\mathbf{u}$ that $\mathbf{u}^T K\mathbf{u} \geq 0$, and equality with zero implies $\mathbf{u} = \mathbf{0}$. We compute

$$\mathbf{u}^T K\mathbf{u} = \sum_{i=1}^{n}\sum_{j=1}^{n} u_i(\varphi_i', \varphi_j')u_j = \left(\sum_{i=1}^{n} u_i\varphi_i', \sum_{j=1}^{n} u_j\varphi_j'\right) = (u_h', u_h') = ||u_h'||^2 \geq 0$$

for the function $u_h(x) := \sum_{j=1}^{n} u_j\varphi_j$. Now if $||u_h'|| = 0$, then the derivative of u_h must vanish identically, $u_h' = 0$, which means that $u_h(x)$ must be the constant function. Since $u_h \in V_h$, $u_h(0) = u_h(1) = 0$, which means that the constant function u_h must vanish identically, and hence K is indeed positive definite. The proof for the mass matrix M is similar. □

Instead of using the weak form (5.4), as Galerkin did in 1915, Ritz originally used in 1908 the equivalent *minimization problem* (5.6) with functions only from the approximate subspace V_h in (5.9). We then obtain the Ritz approximation.

Definition 5.7 (Ritz approximation).

$$\text{Find } u_h \in V_h \text{ such that } F(u_h) \text{ from (5.5) is minimized.} \tag{5.15}$$

Writing u_h again as a linear combination of the functions in V_h,

$$u_h = \sum_{j=1}^{n} u_j\varphi_j,$$

the minimization problem (5.15) turns into the finite dimensional minimization problem

$$F(u_h) = \frac{1}{2}(u_h', u_h') - (f, u_h) = \frac{1}{2}\left(\sum_{i=1}^n u_i\varphi_i', \sum_{j=1}^n u_j\varphi_j'\right) - \left(f, \sum_{i=1}^n u_i\varphi_i\right)$$
$$= \frac{1}{2}\sum_{i=1}^n\sum_{j=1}^n u_i(\varphi_i', \varphi_j')u_j - \sum_{i=1}^n u_i(f, \varphi_i) \quad \longrightarrow \quad \min,$$

and we recognize that this is just a finite dimensional quadratic form to be minimized,

$$\frac{1}{2}\mathbf{u}^T K\mathbf{u} - \mathbf{u}^T\mathbf{f} \quad \longrightarrow \quad \min, \tag{5.16}$$

with the same *stiffness matrix* and right-hand-side function as in the Galerkin formulation, $K_{ij} := (\varphi_i', \varphi_j')$ and $f_i := (f, \varphi_i)$.

A necessary condition for $\mathbf{u}$ to be a minimizer of the quadratic form in (5.16) is that the derivative with respect to $\mathbf{u}$ must vanish,

$$\nabla_{\mathbf{u}}\left(\frac{1}{2}\mathbf{u}^T K\mathbf{u} - \mathbf{u}^T\mathbf{f}\right) = \frac{1}{2}K\mathbf{u} + \frac{1}{2}K^T\mathbf{u} - \mathbf{f} = K\mathbf{u} - \mathbf{f} = 0, \tag{5.17}$$

where we used the fact that K is symmetric; see Lemma 5.6. Hence, the necessary condition for a minimum gives the same linear system as the Galerkin approximation, $K\mathbf{u} = \mathbf{f}$. To see why this condition is also sufficient, we compute the second derivative by taking a derivative of (5.17), and we find for the Hessian the matrix K, which is positive definite by Lemma 5.6. Solving the approximate minimization problem (5.16) of Ritz is thus equivalent to solving the Galerkin approximation system $K\mathbf{u} = \mathbf{f}$, which illustrates the equivalence in Theorem 5.3 also at the discrete level. This approach of finding an approximate solution is nowadays called the Ritz–Galerkin method.

5.3 ▪ More General Boundary Conditions

We had for simplicity assumed in our model problem (5.1) that the boundary conditions are homogeneous Dirichlet conditions. We now consider a more general case with inhomogeneous Dirichlet and Neumann boundary conditions, namely, the strong form

$$\left\{\begin{array}{rcl} -u'' &=& f \quad \text{in } \Omega = (0,1), \\ u(0) &=& g_L, \\ u'(1) &=& g_R. \end{array}\right. \tag{5.18}$$

To find the variational formulation, we multiply again by a test function v and integrate by parts,

$$-\int_0^1 u''(x)v(x)dx = -u'(x)v(x)\Big|_0^1 + \int_0^1 u'(x)v'(x)dx = \int_0^1 f(x)v(x)dx. \tag{5.19}$$

Note that the Neumann condition gives the value of $u'(1) = g_R$, and this value appears in the variational formulation (5.19) from the boundary term after integrating by parts. Thus, in order to keep this information in the variational formulation, one must not impose $v(1) = 0$ in the test space; instead, the test function should only vanish at $x = 0$, $v(0) = 0$, and the value $v(1)$ should be left arbitrary, in contrast to (5.2). In addition, we know that the solution satisfies $u(0) = g_L$, so the set of trial functions in which the solution is sought needs to enforce this explicitly. If we denote by $H^1(\Omega)$ the space of trial functions with no constraints on boundary values,[23] then the weak

[23] This space will be defined precisely in section 5.4.

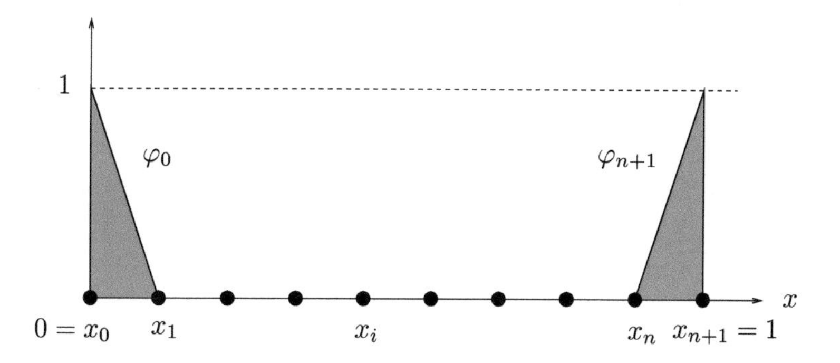

Figure 5.3. *Hat functions φ_0 and φ_{n+1} to include inhomogeneous Dirichlet and Neumann boundary conditions in the finite element formulation.*

or variational form in the case of inhomogeneous boundary conditions and Neumann conditions can be written in inner product notation as

$$\begin{aligned}&\text{find } u \in \{H^1(\Omega), u(0) = g_L\} \text{ such that}\\&(u', v') = (f, v) + v(1)g_R \quad \forall v \in \{H^1(\Omega), v(0) = 0\}.\end{aligned} \tag{5.20}$$

This shows why the Dirichlet condition is called an *essential boundary condition* in the finite element community: It has to be imposed explicitly, with the corresponding boundary value in the set of trial functions and with value equal to zero in the set of test functions. The Neumann condition, on the other hand, is automatically included in the variational form and is thus called a *natural boundary condition* in the finite element community.

Example 5.8. If we use again the hat functions from Example 5.5 to approximate the weak formulation (5.20) with inhomogeneous Dirichlet conditions together with boundary conditions of Neumann type, the approximate solution we seek would be of the form

$$u_h = u_L\varphi_0 + \sum_{j=1}^{n+1} \alpha_j\varphi_j,$$

where φ_0 is the piecewise linear function equal to 1 at $x = 0$ and 0 at all the other mesh points and φ_{n+1} is the additional half-hat function needed for the Neumann condition at $x = 1$; see Figure 5.3. The associated linear system would become

$$K\mathbf{u} = \mathbf{f} + \begin{pmatrix} -u_L(\varphi_0', \varphi_1') \\ \mathbf{0} \\ g_R \end{pmatrix},$$

where the stiffness matrix K is now of size $(n+1) \times (n+1)$, one bigger than with the Dirichlet condition on the right.

5.4 ▪ Sobolev Spaces

Before analyzing the method and considering its generalizations, we need to introduce Sobolev spaces, which underpin much of the theory of elliptic PDEs and finite element methods. The interested reader should refer to [16] for a detailed discussion of such spaces.

Definition 5.9. *Let Ω be an open, bounded, simply connected subset of $\mathbb{R}^d$ ($d \geq 1$). The space $L^2(\Omega)$ consists of the set of Lebesgue measurable functions $u : \Omega \to \mathbb{R}$ whose L^2-norm $\|u\|_{L^2(\Omega)}$ is bounded, i.e.,*

$$\|u\|_{L^2(\Omega)} := \left(\int_\Omega |u(x)|^2 \, dx \right)^{1/2} < \infty.$$

Next, we introduce the notion of weak derivatives. Let φ be a *test function*, i.e., an infinitely differentiable function whose support is contained in a compact subset of Ω. Then whenever u is continuously differentiable, we have the integration by parts formula

$$\int_\Omega \frac{\partial u}{\partial x_i} \varphi \, dx = -\int_\Omega u \frac{\partial \varphi}{\partial x_i} \, dx, \qquad i = 1, \dots, d.$$

Note that the boundary terms that usually arise from integration by parts vanish because $\varphi = 0$ on $\partial\Omega$.

In the case where u may not be differentiable, we say that a measurable function v is the *weak partial derivative* of u with respect to x_i if we have, for all test functions φ,

$$\int_\Omega v\varphi \, dx = -\int_\Omega u \frac{\partial \varphi}{\partial x_i} \, dx.$$

This can be generalized for higher-order derivatives: for a multi-index $\alpha = (\alpha_1, \dots, \alpha_d)$, the weak αth partial derivative of u exists if there is a function v such that

$$\int_\Omega v\varphi \, dx = (-1)^{|\alpha|} \int_\Omega u \frac{\partial^{|\alpha|} \varphi}{\partial x_1^{\alpha_1} \cdots \partial x_d^{\alpha_d}} \, dx$$

for all test functions φ, where $|\alpha| = \alpha_1 + \cdots + \alpha_d$. In this case, we write $v = D^\alpha u$.

Definition 5.10. *Let $k > 0$ be an integer. Then the* Sobolev space *$H^k(\Omega)$ consists of all functions $u \in L^2(\Omega)$ such that for all $|\alpha| \leq k$, $D^\alpha u$ exists and is in $L^2(\Omega)$.*

Example 5.11. Let $\Omega = (0, 1) \subset \mathbb{R}$. If u is a continuous, piecewise linear function, then $u \in L^2(\Omega)$, and u' is a piecewise constant function that is also in $L^2(\Omega)$. Thus, $u \in H^1(\Omega)$. This shows that the space V_h defined in (5.9) is in fact a *subspace* of $H^1(\Omega)$.

Remark 5.1. *If Ω is an open interval in one dimension and $u \in H^1(\Omega)$, one can show that u is absolutely continuous and differentiable almost everywhere in Ω, and its weak derivative coincides with its derivative in the classical sense. However, in higher dimensions, functions in H^1 need not even be continuous, and the classical and weak derivatives do not necessarily coincide.*

Next, we define the subspace $H_0^k(\Omega)$, which is useful when dealing with Dirichlet boundary conditions.

Definition 5.12. *Let $C_c^\infty(\Omega)$ be the set of infinitely differentiable functions whose support is a compact subset of Ω. We define the subspace $H_0^k(\Omega)$ to be the closure of $C_c^\infty(\Omega)$ in $H^k(\Omega)$, so that $f \in H_0^k(\Omega)$ if and only if there is a sequence $(f_k)_{k=1}^\infty$, $f_k \in C_c^\infty(\Omega)$, such that $\|f_k - f\|_{H^k(\Omega)} \to 0$ as $k \to \infty$.*

Roughly speaking, $H_0^k(\Omega)$ consists of functions in $H^k(\Omega)$ for which "$D^\alpha u = 0$ on $\partial\Omega$" for all $|\alpha| \le k-1$. The reason we cannot use this as the definition is because u is not even continuous in general, so we cannot speak of the value of $D^\alpha u$ at a boundary point for a general function $u \in H^k(\Omega)$. To get around this problem, we instead consider the set of all functions in $H^k(\Omega)$ that are "well approximated" by smooth functions that vanish on the boundary, i.e., the set of all functions in $H^k(\Omega)$ that are limits of sequences in $C_c^\infty(\Omega)$. This set is naturally closed and is the smallest closed subset of $H^k(\Omega)$ that contains $C_c^\infty(\Omega)$, which is why it is called the *closure* of $C_c^\infty(\Omega)$ in $H^k(\Omega)$.

For second-order PDEs, the most commonly used spaces are $L^2(\Omega)$, $H^1(\Omega)$, and $H_0^1(\Omega)$. It can be shown that they are Hilbert spaces (i.e., complete inner product spaces) when equipped with the inner products[24]

$$(u,v)_{L^2(\Omega)} = \int_\Omega u(x)v(x)\,dx, \qquad (u,v)_{H^1(\Omega)} = \int_\Omega (u(x)v(x) + \nabla u(x)\cdot\nabla v(x))\,dx,$$

with the latter being valid for both $H^1(\Omega)$ and $H_0^1(\Omega)$. Moreover, this inner product induces the following norm on $H^1(\Omega)$ and $H_0^1(\Omega)$:

$$\|u\|_{H^1(\Omega)} := \left(\int_\Omega u^2(x) + |\nabla u(x)|^2\,dx\right)^{1/2}.$$

For $H_0^1(\Omega)$, it turns out one can use a different norm $\|\cdot\|_a$, defined using only the gradient:

$$\|u\|_a := \left(\int_\Omega |\nabla u(x)|^2\,dx\right)^{1/2}.$$

In fact, the two norms are *equivalent* in the sense that there exist positive constants c and C such that

$$c\|u\|_{H^1(\Omega)} \le \|u\|_a \le C\|u\|_{H^1(\Omega)} \qquad \forall u \in H_0^1(\Omega).$$

This result is a consequence of the very important *Poincaré inequality*, which we state below.

Theorem 5.13 (Poincaré inequality). *Suppose Ω is a bounded, open subset of $\mathbb{R}^d$. Then there exists a constant C, depending only on Ω, such that*

$$\|u\|_{L^2(\Omega)} \le C\|\nabla u\|_{L^2(\Omega)} \qquad \forall u \in H_0^1(\Omega). \tag{5.21}$$

Note that the requirement $u \in H_0^1(\Omega)$ is essential; an inequality such as (5.21) cannot hold for general H^1 functions since we can simply take as a counterexample the constant function $u \equiv 1$, whose L^2 norm is nonzero, but $\nabla u = 0$ identically. To avoid this complication, we need to consider functions of *zero mean*, i.e., H^1 functions v such that $\int_\Omega v = 0$. If we assume in addition that Ω has a smooth (at least C^1) boundary, then for an arbitrary $u \in H^1(\Omega)$ we have

$$\|u - \bar{u}\|_{L^2(\Omega)} \le C\|\nabla u\|_{L^2(\Omega)}, \tag{5.22}$$

where $\bar{u} = \frac{1}{|\Omega|}\int_\Omega u(x)\,dx$ is the mean value of u in the domain Ω, so that $u - \bar{u}$ has zero mean. The proof of (5.21) and (5.22) can be found in [2]. Instead of proving

[24]From now on, we will make no the distinction between classical and weak gradients, as we will only work with functions for which the two concepts coincide.

Theorem 5.13 in its generality, we illustrate it by deriving an estimate in the one-dimensional case when Ω is an interval of length H, which we can take to be $(0, H)$ without loss of generality. Since $u \in H_0^1(\Omega)$, we must have $u(0) = 0$, so that $u(x) = \int_0^x u'(\xi)\, d\xi$ for all $x \in \Omega$. Using the Cauchy–Schwarz inequality, we obtain

$$|u(x)|^2 = \left| \int_0^x 1 \cdot u'(\xi)\, d\xi \right|^2 \leq \left| \int_0^x 1\, d\xi \right| \int_0^x |u'(\xi)|^2\, d\xi \leq x \int_0^H |u'(\xi)|^2\, d\xi.$$

Integrating again with respect to x yields

$$\int_0^H |u(x)|^2\, dx \leq \frac{H^2}{2} \int_0^H |u'(\xi)|^2\, d\xi, \tag{5.23}$$

so (5.21) is satisfied with $C = H/\sqrt{2}$.

We have seen in section 5.2 that the discrete Galerkin problem (5.10) has a unique solution because the stiffness matrix K is symmetric and positive definite; i.e., we have $\mathbf{u}^T K \mathbf{u} > 0$ whenever $\mathbf{u} \neq 0$. In the continuous setting, we need a similar property to ensure that the weak form (5.4) has a unique solution.

Definition 5.14. *Let $a(\cdot, \cdot)$ be a bilinear form on a Hilbert space $\mathcal{H}$, i.e., a mapping of the form $\mathcal{H} \times \mathcal{H} \to \mathbb{R}$ such that $a(\lambda u + \mu v, w) = \lambda a(u, w) + \mu a(v, w)$ and $a(u, \lambda v + \mu w) = \lambda a(u, v) + \mu a(u, w)$ for all $u, v, w \in \mathcal{H}$, $\lambda, \mu \in \mathbb{R}$. Then*

1. *$a(\cdot, \cdot)$ is* bounded *if there exists a constant $\alpha > 0$ such that*

$$a(u, v) \leq \alpha \|u\|\, \|v\| \qquad \forall u, v \in \mathcal{H};$$

2. *$a(\cdot, \cdot)$ is* coercive *if there exists a constant $\beta > 0$ such that*

$$a(u, u) \geq \beta \|u\|^2 \qquad \forall u \in \mathcal{H}.$$

Note that $a(\cdot, \cdot)$ need not be symmetric; i.e., we may have $a(u, v) \neq a(v, u)$. The following theorem is central to the proof of existence and uniqueness of solutions.

Theorem 5.15 (Lax–Milgram). *Let $a(\cdot, \cdot)$ be a bounded, coercive bilinear form and $\ell : \mathcal{H} \to \mathbb{R}$ be a bounded linear functional. Then there exists a unique element $u \in \mathcal{H}$ such that*

$$a(u, v) = \ell(v) \qquad \forall v \in \mathcal{H}.$$

For a proof of this theorem, see [16]. We now illustrate the power of this theorem by proving that the weak form (5.4) of our model problem has a unique solution.

Theorem 5.16. *Let $f \in L^2(0, 1)$. Then there exists a unique function $u \in H_0^1(0, 1)$ such that*

$$(u', v') = (f, v) \qquad \forall v \in H_0^1(0, 1).$$

Proof. Let $a(u, v) = (u', v')$ and $\ell(v) = (f, v)$. Then $a(\cdot, \cdot)$ is bounded since

$$\begin{aligned} a(u, v) = \int_0^1 u'(x) v'(x)\, dx &\leq \left(\int_0^1 |u'(x)|^2\, dx \right)^{1/2} \left(\int_0^1 |v'(x)|^2\, dx \right)^{1/2} \\ &\leq \|u\|_{H^1(0,1)}\, \|v\|_{H^1(0,1)} < \infty \end{aligned}$$

since $u, v \in H^1(0,1)$. Moreover, by the Poincaré inequality (5.23), we have $\|u\|_{L^2(0,1)} \leq \frac{1}{\sqrt{2}}\|u'\|_{L^2(0,1)}$ for all $u \in H_0^1(0,1)$, so that

$$\|u\|_{H^1(0,1)}^2 = \|u\|_{L^2(0,1)}^2 + \|u'\|_{L^2(0,1)}^2 \leq \frac{3}{2}a(u,u).$$

Thus, $a(\cdot,\cdot)$ is coercive with constant $\beta = \frac{2}{3}$. Finally, the linear functional $\ell(v) := (f, v)$ is bounded since the Cauchy–Schwarz inequality implies that

$$|\ell(v)| = |(f,v)| \leq \|f\|_{L^2(0,1)}\|v\|_{L^2(0,1)},$$

so the functional is bounded with constant $\|f\|_{L^2(0,1)}$. Thus, the Lax–Milgram theorem implies that the problem (5.4) has a unique solution $u \in H_0^1(0,1)$, as required. □

Remark 5.2. *Note that $a(\cdot,\cdot)$ is only coercive in $H_0^1(\Omega)$ but not in $H^1(\Omega)$ since $a(u,u) = 0$ for any constant function $u \neq 0$. Thus, the existence and uniqueness result does not hold if we let $V = H^1(\Omega)$ instead of $V = H_0^1(\Omega)$. Indeed, if we have homogeneous Neumann boundary conditions at both $x = 0$ and $x = 1$, then a solution only exists if $\int_0^1 f(x)\,dx = 0$, and it is not unique because one can always add an arbitrary constant to obtain another solution; see also Remark* 4.1 *in Chapter* 4.

5.5 ▪ Convergence Analysis

We now present a convergence analysis of the finite element method for a one-dimensional model problem,

$$\begin{cases} -u'' + u = f & \text{on } \Omega = (0,1), \\ u(0) = u(1) = 0. \end{cases} \tag{5.24}$$

The one-dimensional setting allows us to present all the important ingredients without resorting to overly technical calculations. A similar proof can be derived in higher dimensions, but the required estimates become more involved; see [2].

The weak form of (5.24) can be written as

$$\text{find } u \in V \quad \text{such that} \quad a(u,v) = (f,v) \quad \forall v \in V, \tag{5.25}$$

where $(\cdot,\cdot) = (\cdot,\cdot)_{L^2}$ and the bilinear form $a(u,v)$ is obtained through integration by parts,

$$a(u,v) = \int_0^1 (u(x)v(x) + u'(x)v'(x))\,dx.$$

Note that $a(u,v)$ coincides with $(u,v)_{H^1}$ in this case;[25] if $u, v \in H_0^1(0,1)$, then

$$(u,v)_{H^1} \leq \|u\|_{H^1}\|v\|_{H^1} < \infty.$$

Thus, $H_0^1(0,1)$ is the largest space in which the variational formulation make sense, so an appropriate choice for the *test function space* V is $V = H_0^1(0,1)$.

In order to obtain a finite element method, we divide $\Omega = (0,1)$ into subintervals $x_0 = 0, x_1, \ldots, x_N = 1$, and choose for the *test function space* V_h the space of continuous, piecewise linear function which vanish on the boundary. This leads to the *Galerkin approximation*

$$\text{find } u_h \in V_h \quad \text{such that} \quad a(u_h, v_h) = (f, v_h) \quad \forall v_h \in V_h. \tag{5.26}$$

[25] Since we always have $\Omega = (0,1)$ in this section, we will omit the domain $(0,1)$ in the subscripts and abbreviate $(\cdot,\cdot)_{H^1(0,1)}$ as $(\cdot,\cdot)_{H^1}$, etc.

Since V_h is a subspace of V, the boundedness and coercivity conditions on $a(\cdot,\cdot)$ hold trivially in V_h, so the Lax–Milgram theorem (Theorem 5.15) ensures the existence and uniqueness of the discrete solution u_h of (5.26). We would like to analyze the convergence of u_h to the continuous solution u as $h \to 0$, i.e., as the mesh is refined.

The convergence analysis of a finite element method generally consists of the following three steps:

1. a best approximation result (*Céa's lemma*);
2. an *interpolation estimate*;
3. a duality argument (the *Aubin–Nitsche trick*).

We now show in detail these three mathematical results for our model problem.

Lemma 5.17 (Céa's lemma). *Let u be the weak solution and u_h the solution of the Galerkin approximation. Then the approximation u_h is optimal in the H^1 norm,*

$$\|u - u_h\|_{H^1} \le \|u - w_h\|_{H^1} \quad \forall w_h \in V_h. \tag{5.27}$$

Proof. The inequality is obviously true if $u = u_h$, so we only consider the case where $u \neq u_h$. Since $V_h \subset V = H_0^1$, we obtain from (5.25)

$$a(u, v_h) = (f, v_h) \quad \forall v_h \in V_h,$$

and the Galerkin approximation gives

$$a(u_h, v_h) = (f, v_h) \quad \forall v_h \in V_h.$$

Taking the difference of these two equations yields

$$a(u - u_h, v_h) = 0 \quad \forall v_h \in V_h.$$

But $a(\cdot,\cdot)$ coincides with the H^1 inner product, so we have shown that the error $u - u_h$ is orthogonal to V_h in this inner product. In particular, since $u_h - w_h \in V_h$ whenever $w_h \in V_h$, we have $(u - u_h, u_h - w_h) = 0$ for all $w_h \in V_h$. Hence, for any $w_h \in V_h$, we have

$$\begin{aligned}
\|u - u_h\|_{H^1}^2 &= (u - u_h, u - u_h)_{H^1} \\
&= (u - u_h, u - w_h)_{H^1} - (u - u_h, u_h - w_h)_{H^1} \\
&= (u - u_h, u - w_h)_{H^1} \\
&\le \|u - u_h\|_{H^1} \|u - w_h\|_{H^1},
\end{aligned}$$

where we used the Cauchy–Schwarz inequality in the last step. Dividing both sides by $\|u - u_h\|_{H^1}$ (which is nonzero by our assumption that $u \neq u_h$), we obtain (5.27). □

Remark 5.3. *For a general bilinear form $a(\cdot,\cdot)$ that does not coincide with the H^1 inner product, a similar argument can be used to show that*

$$\|u - u_h\|_{H^1} \le \frac{\alpha}{\beta} \|u - w_h\|_{H^1},$$

where the constants α and β are as in Definition 5.14.

Next, we turn our attention to the interpolation estimate. The key result is that the error between a function u and its piecewise linear interpolant u_I depends on the mesh size h and the second derivative $\|u''\|_{L^2}$. However, we first need to show that u'' exists almost everywhere, which is not obvious because the Lax–Milgram theorem only tells us that $u \in H_0^1(\Omega)$. The following lemma shows this.

Lemma 5.18 (higher regularity). *Let u be the weak solution of $-u'' + u = f$ on $\Omega = (0,1)$, $u(0) = u(1) = 0$ with f continuous. Then u' is differentiable almost everywhere, and $u'' = u - f$ almost everywhere.*

Proof. Let $w = u - f$, which is continuous because f is continuous by assumption and $u \in H_0^1(\Omega)$ is absolutely continuous (see Remark 5.1). Then for all test functions $\varphi \in C_c^\infty(\Omega)$, we have

$$(w, \varphi) = (u, \varphi) - (f, \varphi) = (u, \varphi) - a(u, \varphi) = -(u', \varphi').$$

Thus, w is the weak derivative of u', so $u' \in H^1(\Omega)$. Referring again to Remark 5.1, we see that u' is differentiable almost everywhere in Ω, with its derivative given almost everywhere by the continuous function $u'' = w = u - f$. □

Remark 5.4. *The above argument uses the fact that Ω is one-dimensional in a very essential way. The proof in higher dimensions is much more involved; see* [2].

From now on, we identify u'' with its continuous version. Next, we prove the interpolation estimate.

Lemma 5.19 (interpolation estimate). *Let u be the weak solution of $-u'' + u = f$ on $\Omega = (0,1)$, $u(0) = u(1) = 0$ with f continuous. If u_I is the piecewise linear interpolant of u on the grid $\{x_j\}_{j=0}^N$ with $h = \max|x_j - x_{j-1}|$, then there exists a constant $C > 0$ independent of h and f such that*

$$\|u - u_I\|_{H^1} \leq Ch\|f\|_{L^2}.$$

Proof. Assume again that $u \neq u_h$; otherwise, there is nothing to prove. By definition, we have

$$\|u - u_h\|_{H^1(0,1)}^2 = \int_0^1 |u'(x) - u_I'(x)|^2\,dx + \int_0^1 |u(x) - u_I(x)|^2\,dx. \tag{5.28}$$

We first bound the derivative term. We have

$$\int_0^1 |u'(x) - u_I'(x)|^2\,dx = \sum_{i=1}^N \int_{x_{i-1}}^{x_i} |u'(x) - u_I'(x)|^2\,dx.$$

Since the integrand of each term is differentiable, we can integrate by parts to get

$$\begin{aligned}\int_{x_{i-1}}^{x_i} |u'(x) - u_I'(x)|^2\,dx &= (u - u_I)(u' - u_I')\big|_{x_{i-1}}^{x_i} \\ &\quad - \int_{x_{i-1}}^{x_i} (u(x) - u_I(x))(u''(x) - u_I''(x))\,dx.\end{aligned}$$

The boundary terms vanish because u and u_I coincide there. We also see that $u_I'' = 0$ because u_I is linear on $[x_{i-1}, x_i]$. Thus, we are left with

$$\int_{x_{i-1}}^{x_i} |u'(x) - u_I'(x)|^2 \, dx = -\int_{x_{i-1}}^{x_i} (u(x) - u_I(x))u''(x) \, dx.$$

Summing over all i gives

$$\|u' - u_I'\|_{L^2}^2 = -\int_0^1 (u(x) - u_I(x))u''(x) \, dx,$$

to which we can apply the Cauchy–Schwarz inequality to obtain

$$\|u' - u_I'\|_{L^2}^2 \leq \|u - u_I\|_{L^2} \|u''\|_{L^2}. \tag{5.29}$$

Next, we estimate the L^2 norm of $u - u_I$. Since $u - u_I$ is in H^1 and vanishes at x_{i-1} and x_i, we can use the one-dimensional version of the Poincaré inequality (5.23) to deduce

$$\int_{x_{i-1}}^{x_i} |u(x) - u_I(x)|^2 \, dx \leq \frac{h^2}{2} \int_{x_{i-1}}^{x_i} |u'(x) - u_I'(x)|^2 \, dx.$$

We now sum over all intervals to obtain

$$\|u - u_I\|_{L^2}^2 \leq \frac{h^2}{2} \|u' - u_I'\|_{L^2}^2. \tag{5.30}$$

Combining (5.29) and (5.30), we get

$$\|u - u_I\|_{L^2}^2 \leq \frac{h^2}{2} \|u' - u_I'\|_{L^2}^2 \leq \frac{h^2}{2} \|u - u_I\|_{L^2} \|u''\|_{L^2}.$$

Dividing both sides by $\|u - u_I\|_{L^2(0,1)}$, which is again nonzero by assumption, gives the interpolation estimate

$$\|u - u_I\|_{L^2} \leq \frac{h^2}{2} \|u''\|_{L^2}. \tag{5.31}$$

By reintroducing (5.31) into (5.29), we get

$$\|u' - u_I'\|_{L^2}^2 \leq \frac{h^2}{2} \|u''\|_{L^2}^2. \tag{5.32}$$

Thus, the definition of the H^1 norm gives

$$\begin{aligned} \|u - u_I\|_{H^1}^2 &= \|u' - u_I'\|_{L^2}^2 + \|u - u_I\|_{L^2}^2 \\ &\leq \left(\frac{h^4}{4} + \frac{h^2}{2} \right) \|u''\|_{L^2}^2 \leq Ch^2 \|u''\|_{L^2}^2. \end{aligned} \tag{5.33}$$

Now using the fact that

$$\begin{aligned} \int_0^1 f^2 = \int_0^1 (-u'' + u)^2 &= \|u''\|_{L^2}^2 + \|u\|_{L^2}^2 - 2\int_0^1 u''u \\ &= \|u''\|_{L^2}^2 + \|u\|_{L^2}^2 + 2\|u'\|_{L^2}^2, \end{aligned} \tag{5.34}$$

we conclude that $\|u''\|_{L^2} \leq \|f\|_{L^2}$, and the result follows from (5.33). □

Finally, we prove the convergence estimate using a duality argument due to Aubin and Nitsche.

Lemma 5.20 (Aubin–Nitsche). *Let u be the solution of the variational formulation and u_h the solution of the Galerkin approximation. Then*

$$\|u - u_h\|_{L^2} \leq Ch^2 \|f\|_{L^2}. \tag{5.35}$$

Proof. Assume again that $u \neq u_h$. Let φ be the solution of the auxiliary problem $-\varphi'' + \varphi = u - u_h$, $\varphi(0) = \varphi(1) = 0$, which has the error on the right-hand side. Since $u - u_h$ is continuous, the auxiliary problem has the same structure as the original problem, so we can mimic the calculation in (5.34) to obtain

$$\begin{aligned}\int_0^1 (-\varphi'' + \varphi)^2 dx &= \int_0^1 (\varphi'')^2 - 2\varphi''\varphi + \varphi \\ &= \|\varphi''\|_{L^2}^2 + 2\|\varphi'\|_{L^2}^2 + \|\varphi\|_{L^2}^2 = \|u - u_h\|_{L^2}^2,\end{aligned}$$

which leads to the estimate

$$\|\varphi''\|_{L^2}^2 \leq \|u - u_h\|_{L^2}^2. \tag{5.36}$$

Now the variational formulation of the auxiliary problem is

$$\text{find } \varphi \in H_0^1(\Omega) \text{ such that} \quad a(\varphi, v) = (u - u_h, v) \quad \forall v \in H_0^1(\Omega).$$

Choosing for the test function $v = u - u_h$, we get

$$\|u - u_h\|_{L^2}^2 = a(\varphi, u - u_h) = a(\varphi, u - u_h) - a(\varphi_I, u - u_h) = a(\varphi - \varphi_I, u - u_h),$$

where we used Lemma 5.17. We can now use the Cauchy–Schwarz inequality and (5.36) to obtain

$$\begin{aligned}\|u - u_h\|_{L^2}^2 &\leq \|\varphi - \varphi_I\|_{H^1} \|u - u_h\|_{H^1} \\ &\leq C_1 h \|\varphi''\|_{L^2} \cdot C_2 h \|f\|_{L^2} \leq Ch^2 \|u - u_h\|_{L^2} \|f\|_{L^2},\end{aligned}$$

which leads after division by $\|u - u_h\|_{L^2}$ to the result announced in (5.35). □

5.6 ▪ Generalization to Two Dimensions

We now show how the hat function approximation can be generalized to higher spatial dimensions and consider the two-dimensional model problem in the strong form

$$\left\{ \begin{array}{ll} \eta u - \Delta u = f & \text{in a polygon } \Omega \subset \mathbb{R}^2, \eta \geq 0, \\ u = 0 & \text{on } \partial\Omega. \end{array} \right. \tag{5.37}$$

As in one spatial dimension, we multiply by a test function v and integrate over the domain to find

$$\int_\Omega (\eta uv - \Delta uv) dx = -\int_{\partial\Omega} \frac{\partial u}{\partial n} v ds + \int_\Omega (\eta uv + \nabla u \cdot \nabla v) dx = \int_\Omega f v dx.$$

Defining the bilinear forms

$$a(u,v) := \int_\Omega (\eta uv + \nabla u \cdot \nabla v)dx, \quad (u,v) := \int_\Omega uv dx, \tag{5.38}$$

we see that a natural choice for the function space to search for a weak solution is $V := H_0^1(\Omega)$ since it is the largest space in which the above integrals are well defined. Thus, the weak or variational form is

$$\text{find } u \in H_0^1(\Omega) \text{ such that } a(u,v) = (f,v) \quad \forall v \in H_0^1. \tag{5.39}$$

By the Lax–Milgram theorem, the above problem has a unique solution. Proceeding as in the one-dimensional case, we see that (5.39) is equivalent to the minimization formulation

$$\text{find } u \in H_0^1(\Omega) \text{ such that } F(u) := \tfrac{1}{2}a(u,u) - (f,u) \text{ is minimized.} \tag{5.40}$$

A Ritz–Galerkin approximation is obtained when one replaces the infinite dimensional function space $H_0^1(\Omega)$ by an approximation V_h; thus, we have the *Galerkin approximation*

$$\text{find } u_h \in V_h \text{ such that } a(u_h,v_h) = (f,v_h) \quad \forall v_h \in V_h \tag{5.41}$$

and the *Ritz approximation*

$$\text{find } u_h \in V_h \text{ such that } \tfrac{1}{2}a(u_h,u_h) - (f,u_h) \text{ is minimized,} \tag{5.42}$$

which are also equivalent.

For a typical finite element approximation, one often uses a triangulation of the domain Ω, as we have seen already in the historical examples in Figure 5.1.[26] One then defines the approximation subspace V_h by

$$V_h = \text{span}\{\varphi_1, \ldots, \varphi_n\},$$

where every φ_j is a hat function; i.e., φ_j is an affine function on each triangle, and

$$\varphi_j = \begin{cases} 1 & \text{at } x_j, \\ 0 & \text{at } x_i \neq x_j; \end{cases}$$

see Figure 5.4 for an example. One thus needs a mesh generator to be able to do finite element computations in two space dimensions. A very simple procedure using MATLAB to create some initial meshes is `NewMesh`:

```
function [N,T,P]=NewMesh(G);
% NEWMESH generates a simple new mesh for predefined domains
%   [N,T,P]=NewMesh(G); generates an initial coarse triangular
%   mesh. Use G=0 for the default square, G=1 for a triangle, G=2 for
%   a space shuttle and G=3 for an empty micro wave oven, and G=4 for a
%   chicken in a micro wave oven. The result is a table of triangles T
%   which points into a table of nodes N containing x and y
%   coordinates.  The triangle contains in entries 4 to 6 a 1 if its
%   corresponding edges are real boundaries. P can contain for each
%   triangle a material property.
```

[26] Courant 1941: "Instead of starting with a quadratic or rectangular net we may consider from the outset any polyhedral surfaces with edges over an *arbitrarily chosen* (preferably triangular) *net*."

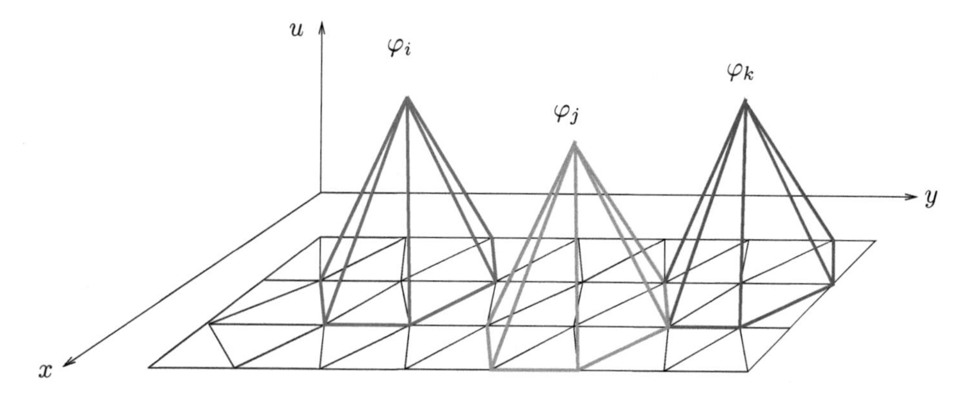

Figure 5.4. *Examples of hat functions in two dimensions.*

```
P=[];                                               % default no material
if G==1,                                            % triangle
  N=[0 0; 1 0; 0.5 1];
  T=[1 2 3 1 1 1];
elseif G==2                                         % space shuttle
  N=[0.07 0; 1 0; 1 0.4; 0.8 0.23; 0.25 0.23; 0.15 0.16; 0.07 0.15; 0 0.1
    0 0.05; 0.93 0.4; 0.5 0; 1 0.1; 1 0.15; 1.12 0.08; 1.12 0.17; 0.15 0
    0.07 0.07; 0.8 0; 0.25 0; 0.9 0.1];
  T=[1 17 9 0 0 1; 9 17 8 0 0 1; 8 17 7 0 0 1; 1 16 17 1 0 0; 17 6 7 0 1 0
     16 6 17 0 0 0; 16 19 6 1 0 0; 19 5 6 0 1 0; 19 11 5 1 0 0; 5 11 4 0 0 1
     11 18 4 1 0 0; 18 20 4 0 0 0; 18 2 20 1 0 0; 2 12 20 1 0 0
     20 12 13 0 0 0; 20 13 4 0 0 0; 4 13 10 0 0 1; 10 13 3 0 1 1;
     12 14 13 1 0 0; 14 15 13 1 1 0];
elseif G==3                                         % empty microwave
  N=[0 0; 5 0; 0 3; 5 3];
  T=[1 2 4 1 1 0; 1 4 3 0 1 1];
  P=ones(1,2);
elseif G==4                                         % chicken in a microwave
  N=[0.8 0.9; 1.4 0.4; 2   0.3; 3   0.3; 3.5 0.4    % inside
     4   1  ; 3.5 1.6; 3   2  ; 2.5 2.2; 2   2.4
     1.4 2.4; 1   2  ; 0.5 2  ; 0.3 2.2; 0.2 1.9
     0.3 1.6; 0.5 1.8; 1   1.8; 1.3 1.4; 1.5 1.8
     2   2  ; 1.5 1  ; 2   1  ; 3   1  ; 3.5 1
     0   0  ; 1   0  ; 2   0  ; 3   0  ; 4   0       % outside
     5   0  ; 5   3  ; 4   3  ; 2.5 3  ; 1.5 3
     0.8 3  ; 0   3  ; 0   1.9; 0   1  ; 5   1.5];
  T=[13 14 15 0 0 0; 15 16 17 0 0 0; 15 17 13 0 0 0; 17 12 13 0 0 0
     17 18 12 0 0 0; 18 20 12 0 0 0; 18 19 20 0 0 0; 12 20 11 0 0 0
     20 21 11 0 0 0; 21 10 11 0 0 0; 21  9 10 0 0 0;  1 22 19 0 0 0
     22 20 19 0 0 0; 22 23 20 0 0 0;  2 22  1 0 0 0;  2 23 22 0 0 0
     20 23 21 0 0 0; 23  9 21 0 0 0;  2  3 23 0 0 0;  3 24 23 0 0 0
     23 24  9 0 0 0;  3  4 24 0 0 0; 24  8  9 0 0 0; 24  7  8 0 0 0
     24 25  7 0 0 0; 24  5 25 0 0 0;  4  5 24 0 0 0;  5  6 25 0 0 0
      6  7 25 0 0 0; 26 27  1 1 0 0; 27  2  1 0 0 0; 27 28  2 1 0 0
     28  3  2 0 0 0; 28 29  3 1 0 0; 29  4  3 0 0 0; 29 30  4 1 0 0
     30  5  4 0 0 0; 30  6  5 0 0 0; 30 31  6 1 0 0; 31 40  6 1 0 0
     32 33 40 1 0 1; 33  7  6 0 0 0; 33  8  7 0 0 0; 33 34  8 1 0 0
     34  9  8 0 0 0; 34 10  9 0 0 0; 34 35 10 1 0 0; 35 11 10 0 0 0
     35 36 11 1 0 0; 36 12 11 0 0 0; 36 13 12 0 0 0; 36 14 13 0 0 0
     36 37 14 1 0 0; 37 38 14 1 0 0; 38 15 14 0 0 0; 38 16 15 0 0 0
     38 39 16 1 0 0; 39  1 16 0 0 0;  1 17 16 0 0 0;  1 18 17 0 0 0
      1 19 18 0 0 0; 39 26  1 1 0 0; 40 33  6 0 0 0];
  P=[5*ones(1,29) ones(1,34)];
```

```
else                                                % default square
  N=[0 0; 1 0; 0 1; 1 1];
  T=[1 2 4 1 1 0; 1 4 3 0 1 1];
end;
```

This code produces a few domains with initial triangulations, and it is very easy to add further domains. For example, for the default square (see Figure 5.5, top left), we obtain the two matrices

$$N = \begin{pmatrix} 0 & 0 \\ 1 & 0 \\ 0 & 1 \\ 1 & 1 \end{pmatrix}, \quad T = \begin{pmatrix} 1 & 2 & 4 & 1 & 1 & 0 \\ 1 & 4 & 3 & 0 & 1 & 1 \end{pmatrix}.$$

N indicates that there are four nodes (x_j, y_j), $j = 1, 2, 3, 4$ in this mesh, namely, $(0, 0)$, $(0, 1)$, $(1, 0)$, $(1, 1)$, and the mesh contains two triangles as indicated in T, one consisting of the nodes 1, 2, and 4 in counterclockwise order and the other one consisting of the nodes 1, 4, and 3, also in counterclockwise order—a consistent ordering is important later when computing on the triangles. The remaining three entries in each triangle indicate which sides of the triangle represent a real boundary of the domain; e.g., $1, 2, 4, 1, 1, 0$ for the first triangle means that the edge linking nodes 1 and 2 is a boundary, indicated by the first 1, and the edge linking nodes 2 and 4 is also a boundary, indicated by the second 1, but the edge linking node 4 to node 1 is an interior edge, indicated by the last 0. Optionally, one may specify in the matrix P a coefficient that is piecewise constant on each triangle; see G=3 and G=4 in the code above. Such a coefficient often represents a physical property that appears in the PDE and can be taken into account later by the finite element code. The resulting initial mesh can be visualized using the function `PlotMesh`:

```
function PlotMesh(N,T,P);
% PLOTMESH plots a triangular mesh
%   PlotMesh(N,T,P); plots the mesh given by the nodes N and triangles
%   T. The real boundaries are drawn in bold and for small meshes
%   the node numbers are added as well. If P is given as an input argument,
%   P contains an element by element physical property

clf; axis('equal');
if nargin==3 & ~isempty(P),                   % display material property
  for i=1:size(T,1),
    patch(N(T(i,1:3),1),N(T(i,1:3),2),P(i));
  end
end
for i=1:size(T,1),                            % plot mesh
  for j=1:3,
    line([N(T(i,j),1) N(T(i,mod(j,3)+1),1)], ...
      [N(T(i,j),2) N(T(i,mod(j,3)+1),2)],'LineWidth',T(i,j+3)*3+1);
  end
end
m=size(N,1);
if m<100,                                     % dislay mesh nodes for
  for i=1:m,                                  % small meshes
    text(N(i,1)+.01,N(i,2)+.02,num2str(i));
  end
end
```

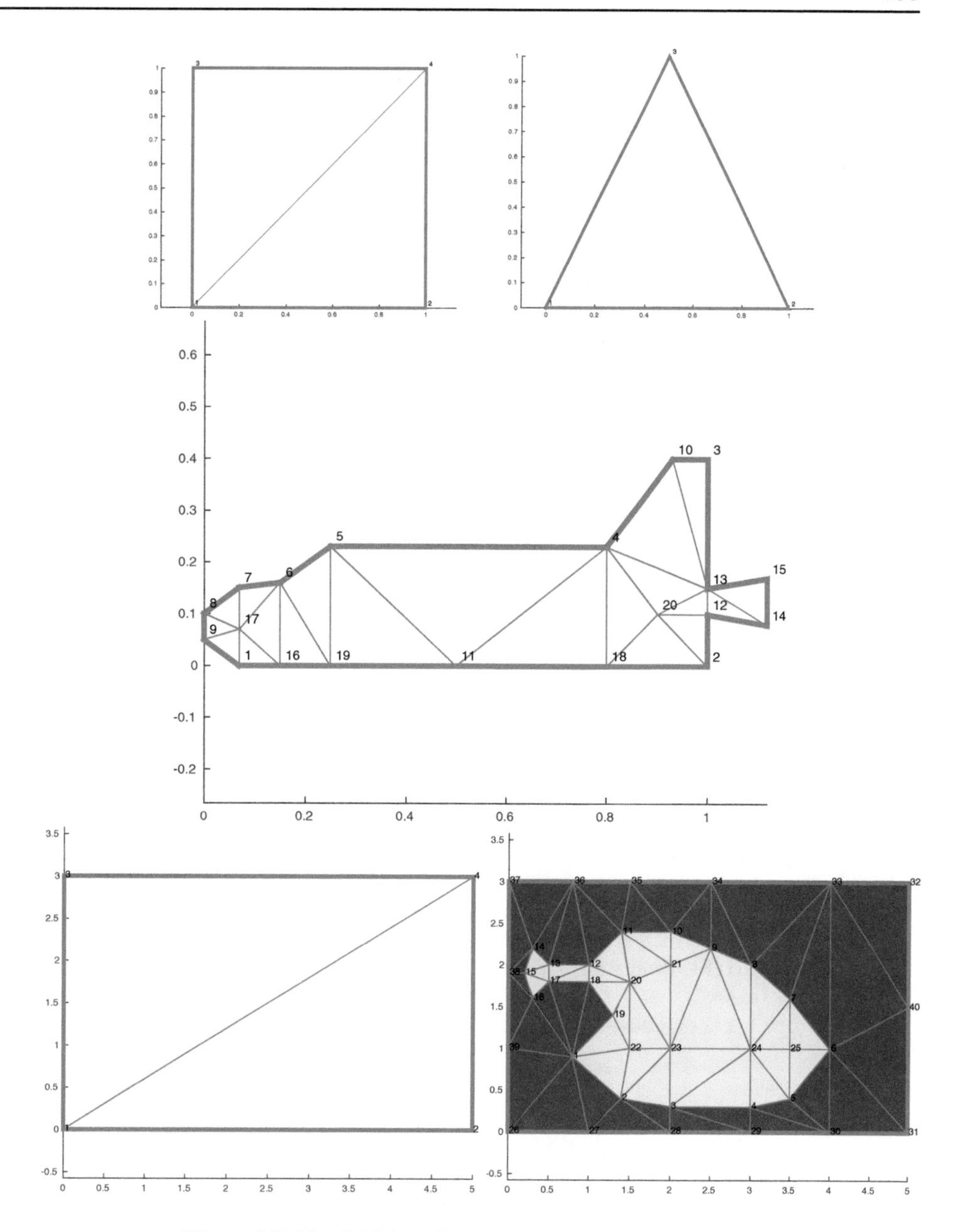

Figure 5.5. *Five initial meshes one can produce using* `NewMesh`.

We show in Figure 5.5 the five initial meshes one can generate with `NewMesh` using the MATLAB commands

```
for i=0:3
  [N,T]=NewMesh(i);
  PlotMesh(N,T);
end
[N,T,P]=NewMesh(4);
PlotMesh(N,T,P);
```

For the last mesh, we also plot the material property using different colors for different values. Once we have an initial mesh, we would often want to refine it to obtain more triangles, which are required for a more accurate result. For meshes generated with `NewMesh`, this can be achieved using the following MATLAB routine `RefineMesh`:

```
function [Nr,Tr,Pr]=RefineMesh(N,T,P);
% REFINEMESH refines the mesh by a factor of 4
%   [Nr,Tr,Pr]=RefineMesh(N,T,P); refines the mesh given by the nodes N and
%   the triangles T and physical property P on each triangle by cutting each
%   triangle into four smaller ones. The boundary is traced so that the new
%   triangles still contain the information that they are touching the
%   boundary.

Nr=N;                              % new node list starts with old one
nn=size(N,1);
Tr=zeros(size(T,1)*4,6);           % triangles start from scratch
nt=0;
if nargin==3
  Pr=zeros(1,length(P)*4);
else
  Pr=[];
end;
NewNid=sparse(nn,nn);
Nid=[1 2;1 3;2 3];
for j=1:size(T,1),                 % quadrisect all old triangles
  i=T(j,1:3); n=N(i,:);            % old nodes of current triangle
  for k=1:3
    i1=Nid(k,1);i2=Nid(k,2);
    if i(i1)>i(i2), tmp=i1; i1=i2; i2=tmp; end;   % to only store once
    if NewNid(i(i1),i(i2))==0
      nn=nn+1; Nr(nn,:)=(n(i1,:)+n(i2,:))/2;
      i(k+3)=nn; NewNid(i(i1),i(i2))=nn;          % store node
    else
      i(k+3)=NewNid(i(i1),i(i2));
    end
  end
  Tr(nt+1,:)=[i(1) i(4) i(5) T(j,4) 0 T(j,6)];   % insert 4 new triangles
  Tr(nt+2,:)=[i(5) i(4) i(6) 0 0 0];
  Tr(nt+3,:)=[i(6) i(4) i(2) 0 T(j,4) T(j,5)];
  Tr(nt+4,:)=[i(6) i(3) i(5) T(j,5) T(j,6) 0];
  if nargin==3
    Pr(nt+1:nt+4)=P(j);
  end
  nt=nt+4;
end;
```

This procedure cuts each triangle into four smaller triangles, keeps track of new nodes that have to be added, and decides whether each new edge is on the physical boundary of the domain. Now refining triangles is often not enough to produce a good-quality mesh, which should be smooth and have triangles that have no obtuse angles. A simple procedure to improve mesh quality is known as *mesh smoothing*: One simply replaces the coordinates of each node by the average of those of the neighboring nodes and repeats the process several times.[27] The procedure SmoothMesh does this:

```
function N=SmoothMesh(N,T,bn);
% SMOOTHMESH smooth mesh closer to improve mesh quality
%   Nr=SmootheMesh(N,T,bn); modify node positions iteratively to get
%   closer to a Delauny triangulation, by iteratively assigning to
%   nodes the averages of neighboring nodes. Node numbers in bn are
%   not allowed to have their position be modified

nn=size(N,1);
Nn=zeros(nn,1);
nt=size(T,1);                       % number of triangles
if nargin<3
  bn=[];
end
for i=1:nt                          % first find boundary nodes
  for j=1:3
    if T(i,3+j)==1 | T(i,mod(3+j+1,3)+4)==1
      if isempty(find(bn==T(i,j)))
        bn=[bn T(i,j)];
      end
    end
  end
end
for i=1:nt                          % construct neigboring node list
  for j=1:3
    if isempty(find(bn==T(i,j))) % not a boundary node
      for k=1:2                     % add neigboring nodes from triangle
        id=find(Nn(T(i,j),2:Nn(T(i,j),1)+1)==T(i,mod(j+k-1,3)+1));
        if isempty(id)
          Nn(T(i,j),1)=Nn(T(i,j),1)+1;
          Nn(T(i,j),Nn(T(i,j),1)+1)=T(i,mod(j+k-1,3)+1);
        end
      end
    end
  end
end
for i=1:10
  for j=1:nn
    if Nn(j,1)>0
      N(j,:)=mean(N(Nn(j,2:Nn(j,1)+1),:));
    end
  end
end
```

Using the MATLAB commands

```
[N,T]=NewMesh(2);
[N,T]=RefineMesh(N,T);
[N,T]=RefineMesh(N,T);
PlotMesh(N,T);
N=SmoothMesh(N,T);
PlotMesh(N,T);
```

[27]This simple procedure can produce invalid meshes on nonconvex domains due to mesh tangling, so it should be used with care!

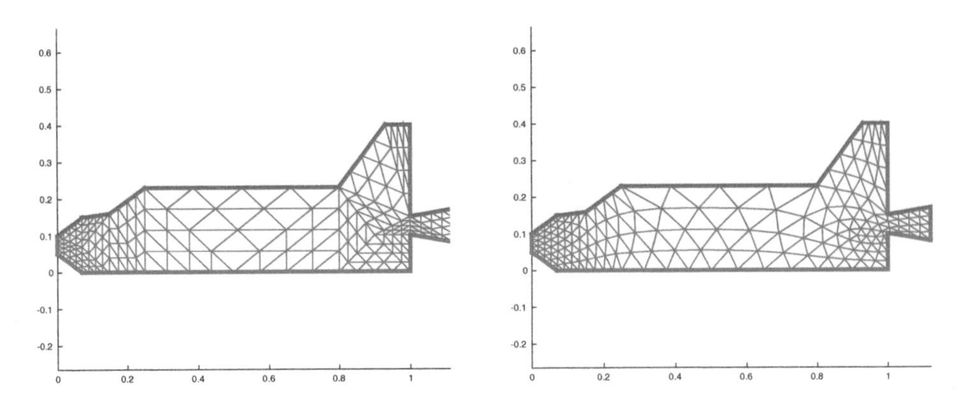

Figure 5.6. *Twice refined mesh of the initial shuttle mesh, before and after mesh smoothing.*

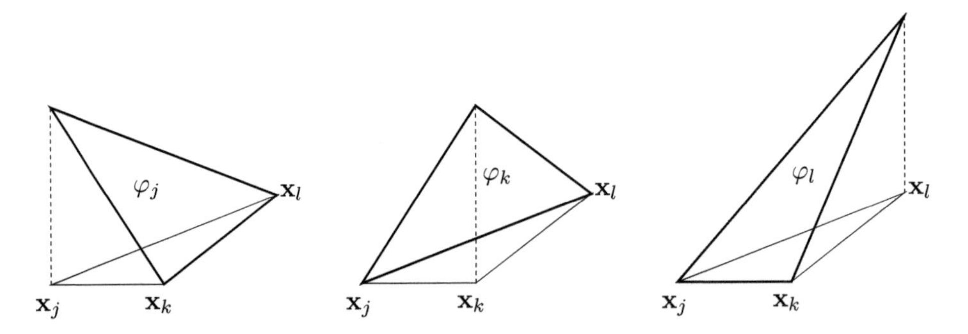

Figure 5.7. *Finite element shape functions on the triangle with vertices* $\mathbf{x}_j$, $\mathbf{x}_k$, $\mathbf{x}_l$.

we obtain the two meshes shown in Figure 5.6. One can clearly see the initial coarse mesh in the refined mesh, before mesh smoothing is applied.

Once a triangular mesh is generated, one can compute the associated stiffness matrix K and mass matrix M in two dimensions, just as in the one-dimensional case. The algorithm for doing so is, however, substantially more complicated than in one dimension, especially when one tries to integrate hat functions against each other for an arbitrary triangular mesh. This is where the true finite elements come in, as we will see next.

5.7 ▪ Where Are the Finite Elements?

Instead of considering the hat functions for computing the *stiffness matrix* K and the *mass matrix* M in a global manner, it is much better to consider the so-called *finite element shape functions* defined on each element (triangle), as illustrated in Figure 5.7. The finite element shape functions are the restrictions of the hat functions to each finite element, and the required integrals for the construction of the stiffness matrix can be computed element by element,

$$\int_\Omega \nabla\varphi_i \cdot \nabla\varphi_j dx = \sum_{k=1}^{n_e} \int_{T_k} \nabla\varphi_i \cdot \nabla\varphi_j dx,$$

where n_e is the number of elements T_k in the triangulation. To simplify the notation in

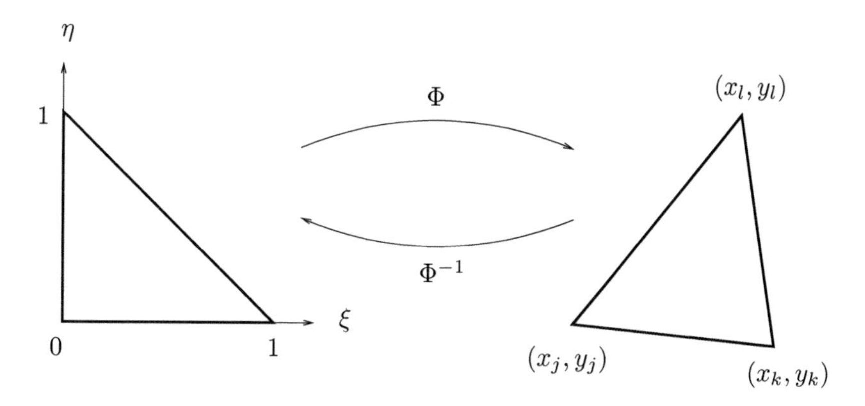

Figure 5.8. *Linear map Φ to transform a reference element T_{ref} into an arbitrary finite element T_i.*

what follows, we introduce the elementwise inner product

$$(\nabla\varphi_i, \nabla\varphi_j)_{T_k} := \int_{T_k} \nabla\varphi_i \cdot \nabla\varphi_j dx.$$

In a *finite element program*, one first computes the *element stiffness matrix* K_i on each element T_i with vertices $\mathbf{x}_j$, $\mathbf{x}_k$ and $\mathbf{x}_l$ by computing

$$K_i := \begin{bmatrix} (\nabla\varphi_j, \nabla\varphi_j)_{T_i} & (\nabla\varphi_j, \nabla\varphi_k)_{T_i} & (\nabla\varphi_j, \nabla\varphi_l)_{T_i} \\ & (\nabla\varphi_k, \nabla\varphi_k)_{T_i} & (\nabla\varphi_k, \nabla\varphi_l)_{T_i} \\ \text{Sym.} & & (\nabla\varphi_l, \nabla\varphi_l)_{T_i} \end{bmatrix} =: \begin{bmatrix} p_{11} & p_{12} & p_{13} \\ & p_{22} & p_{23} \\ \text{Sym.} & & p_{33} \end{bmatrix}.$$

One then adds this element contribution K_i at the appropriate location j, k, l in the *global stiffness matrix* K,

$$K = K + \begin{bmatrix} 0 & & & & & 0 \\ & p_{11} & & p_{12} & & p_{13} \\ & & 0 & & 0 & \\ & p_{12} & & p_{22} & & p_{23} \\ & & 0 & & 0 & \\ & p_{13} & & p_{23} & & p_{33} \\ 0 & & & & & 0 \end{bmatrix} \begin{matrix} \\ j \\ \\ k \\ \\ l \\ \\ \end{matrix}$$
$$\qquad\qquad j \qquad\quad k \qquad\quad l$$

and similarly for the *global mass matrix* M. This is called the *assembly process.*

In order to compute the element stiffness matrix K_i and element mass matrix M_i on the element T_i with vertices $\mathbf{x}_j = (x_j, y_j)$, $\mathbf{x}_k = (x_k, y_k)$, and $\mathbf{x}_l = (x_l, y_l)$, one introduces a linear map Φ from a *reference triangle* T_{ref} to T_i, as illustrated in Figure 5.8. The element shape functions on the reference triangle are very simple:

$$N_1(\xi, \eta) = 1 - \xi - \eta,$$
$$N_2(\xi, \eta) = \xi,$$
$$N_3(\xi, \eta) = \eta.$$

The linear map Φ can then be defined using the element shape functions themselves,[28]

$$\Phi(\xi, \eta) = \begin{pmatrix} x(\xi, \eta) \\ y(\xi, \eta) \end{pmatrix} = \begin{pmatrix} x_j \\ y_j \end{pmatrix} N_1 + \begin{pmatrix} x_k \\ y_k \end{pmatrix} N_2 + \begin{pmatrix} x_l \\ y_l \end{pmatrix} N_3.$$

[28]If this is possible, the element is called an *isoparametric element.*

Using the change of variables formula from multivariate calculus, the integration on every element T_i can be performed on the reference element T_{ref},

$$\int_{T_i} g(\mathbf{x}) d\mathbf{x} = \int_{T_{\text{ref}}} g(\Phi(\xi, \eta)) |J(\xi, \eta)| d\xi d\eta,$$

where $|J(\xi, \eta)| = \det J(\xi, \eta)$ is the determinant of the Jacobian of the map Φ,

$$J(\xi, \eta) = \begin{bmatrix} \frac{\partial x}{\partial \xi} & \frac{\partial x}{\partial \eta} \\ \frac{\partial y}{\partial \xi} & \frac{\partial y}{\partial \eta} \end{bmatrix} = \begin{bmatrix} x_k - x_j & x_l - x_j \\ y_k - y_j & y_l - y_j \end{bmatrix}.$$

To illustrate these computations, we now compute a concrete entry of the element stiffness matrix K_i:

$$\int_{T_i} \nabla \varphi_j \cdot \nabla \varphi_k d\mathbf{x}.$$

To do so, we will need the two relations,

$$\varphi_j(\Phi(\xi, \eta)) = N_1(\xi, \eta) \quad \text{and} \quad \nabla \varphi_j^T J(\xi, \eta) = \left(\frac{\partial N_1}{\partial \xi}, \frac{\partial N_1}{\partial \eta} \right).$$

We then obtain

$$\nabla \varphi_j = J^{-T}(\xi, \eta) \begin{pmatrix} \frac{\partial N_1}{\partial \xi} \\ \frac{\partial N_1}{\partial \eta} \end{pmatrix} = \frac{1}{\det J} \begin{bmatrix} \frac{\partial y}{\partial \eta} & -\frac{\partial y}{\partial \xi} \\ -\frac{\partial x}{\partial \eta} & \frac{\partial x}{\partial \xi} \end{bmatrix} \begin{pmatrix} -1 \\ -1 \end{pmatrix}.$$

We can now compute

$$\begin{aligned} \int_{T_i} \nabla \varphi_j \cdot \nabla \varphi_k d\mathbf{x} &= \int_{T_i} \nabla \varphi_j^T \nabla \varphi_k d\mathbf{x} \\ &= |\det J| (-1, -1) J^{-1} J^{-T} \begin{pmatrix} 1 \\ 0 \end{pmatrix} \underbrace{\int_{T_{\text{ref}}} d\xi d\eta}_{=\frac{1}{2}} \\ &= \frac{1}{2|\det J|} (y_k - y_l, x_l - x_k) \begin{pmatrix} y_l - y_j \\ x_j - x_l \end{pmatrix} \\ &= \frac{(y_k - y_l)(y_l - y_j) - (x_k - x_l)(x_j - x_l)}{2|(x_k - x_j)(y_l - y_j) - (x_l - x_j)(y_k - y_j)|}. \end{aligned}$$

Such cumbersome computations can be performed very easily using Maple. We start by computing the transformation Φ:

```
X:=a+b*xi+c*eta;
Y:=d+e*xi+f*eta;
xi:=0:eta:=0:X1:=X;Y1:=Y;
xi:=1:eta:=0:X2:=X;Y2:=Y;
xi:=0:eta:=1:X3:=X;Y3:=Y;
xi:='xi';
eta:='eta';
solve({X1=x[1],Y1=y[1],X2=x[2],Y2=y[2],X3=x[3],Y3=y[3]},{a,b,c,d,e,f});
assign(%);
```

We then compute the Jacobian and its determinant:

```
with(linalg):
J:=jacobian([X,Y],[xi,eta]);
Jd:=det(J);
Jinv:=inverse(J);
```

Next, we define the local shape functions:

```
l[1]:=1-xi-eta;
l[2]:=xi;
l[3]:=eta;
```

Finally, we compute the resulting formulas for the element stiffness and element mass matrices on a single element:

```
K:=matrix(3,3);
M:=matrix(3,3);
for i from 1 to 3 do
  for j from 1 to 3 do
    M[i,j]:=int(int(l[i]*l[j]*Jd,xi=0..1-eta),eta=0..1);
    gradi:=multiply(transpose(Jinv),grad(l[i],[xi,eta]));
    gradj:=multiply(transpose(Jinv),grad(l[j],[xi,eta]));
    K[i,j]:=int(int(multiply(gradi,gradj)*Jd,xi=0..1-eta),eta=0..1);
  od;
od;
for i from 1 to 3 do
for j from 1 to 3 do
  temp:=simplify(K[i,j]*Jd);
  print(factor(select(has,temp, x)) + factor(select(has,temp, y)));
od;
od;
                                    2                      2
                     1/2 (-x[3] + x[2])  + 1/2 (-y[3] + y[2])
      - 1/2 (-x[3] + x[2]) (x[1] - x[3]) - 1/2 (-y[3] + y[2]) (y[1] - y[3])
       1/2 (-x[3] + x[2]) (x[1] - x[2]) + 1/2 (-y[3] + y[2]) (y[1] - y[2])
      - 1/2 (-x[3] + x[2]) (x[1] - x[3]) - 1/2 (-y[3] + y[2]) (y[1] - y[3])
                                    2                    2
                      1/2 (x[1] - x[3])  + 1/2 (y[1] - y[3])
       - 1/2 (x[1] - x[3]) (x[1] - x[2]) - 1/2 (y[1] - y[3]) (y[1] - y[2])
       1/2 (-x[3] + x[2]) (x[1] - x[2]) + 1/2 (-y[3] + y[2]) (y[1] - y[2])
       - 1/2 (x[1] - x[3]) (x[1] - x[2]) - 1/2 (y[1] - y[3]) (y[1] - y[2])
                                    2                    2
                      1/2 (x[1] - x[2])  + 1/2 (y[1] - y[2])
for i from 1 to 3 do
for j from 1 to 3 do
  print(simplify(M[i,j]/Jd));
od;
od;
                                 1/12
                                 1/24
                                 1/24
                                 1/24
                                 1/12
                                 1/24
```

```
1/24
1/24
1/12
```

Note that the last two double loops are only for formatting purposes, so one can read the results more easily and put them into MATLAB. To compute the element stiffness matrix, the function is

```
function Ke=ComputeElementStiffnessMatrix(t);
% COMPUTEELEMENTSTIFFNESSMATRIX element stiffness matrix for a triangle
%   Ke=ComputeElementStiffnessMatrix(t); computes the element stiffness
%   matrix for the triangular element described by the three nodal
%   coordinates in t=[x1 y1;x2 y2;x3 y3]. Note that the labelling must be
%   counter clockwise

x1=t(1,1); y1=t(1,2);
x2=t(2,1); y2=t(2,2);
x3=t(3,1); y3=t(3,2);

Jd=-x1*y3-x2*y1+x2*y3+x1*y2+x3*y1-x3*y2;          % formulas from Maple
Ke=1/Jd/2*[(x2-x3)^2+(y2-y3)^2 ...
           -(x2-x3)*(x1-x3)-(y2-y3)*(y1-y3) ...
           (x2-x3)*(x1-x2)+(y2-y3)*(y1-y2)
           -(x2-x3)*(x1-x3)-(y2-y3)*(y1-y3) ...
           (x1-x3)^2+(y1-y3)^2 ...
           -(x1-x3)*(x1-x2)-(y1-y3)*(y1-y2)
           (x2-x3)*(x1-x2)+(y2-y3)*(y1-y2) ...
           -(x1-x3)*(x1-x2)-(y1-y3)*(y1-y2) ...
           (x1-x2)^2+(y1-y2)^2];
```

and to compute the element mass matrix, the function is

```
function Me=ComputeElementMassMatrix(t);
% COMPUTEELEMENTMASSMATRIX element mass matrix for a triangle
%   Me=ComputeElementMassMatrix(t); computes the element mass matrix
%   for the triangular element described by the three nodal coordinates
%   in t=[x1 y1;x2 y2;x3 y3], where the nodes are labeled counter clock
%   wise

x1=t(1,1); y1=t(1,2);
x2=t(2,1); y2=t(2,2);
x3=t(3,1); y3=t(3,2);

Jd=-x1*y3-x2*y1+x2*y3+x1*y2+x3*y1-x3*y2;          % formulas from Maple
Me=Jd/24*[2 1 1; 1 2 1; 1 1 2];
```

Remark 5.5. *We used exact integration in this simple example to evaluate the entries of the element stiffness and mass matrices. In practice, this integration is often done using quadrature rules, and one chooses the order of the quadrature rule to still obtain an exact integration since only polynomial functions are integrated. The same quadrature can then also be used to integrate the inner products with the right-hand side, which are then naturally approximated.*

It remains to program the assembly of the element stiffness and element mass matrices, which can be done as follows:

```
function u=FEPoisson(f,g,N,T);
% FEPOISSON solves the Poisson equation using finite elements
%   u=FEPoisson(f,g,N,T); solves the Poisson equation on the
%   triangular mesh given by the list of triangles T and nodes N
%   with Dirichlet boundary conditions given by the function g and
%   forcing function f.

n=size(T,1); m=size(N,1);
bn=zeros(m,1);
K=sparse(m,m); M=sparse(m,m);
for i=1:n,
  Ke=ComputeElementStiffnessMatrix([N(T(i,1),:);N(T(i,2),:);N(T(i,3),:)]);
  Me=ComputeElementMassMatrix([N(T(i,1),:); N(T(i,2),:); N(T(i,3),:)]);
  bn(T(i,1))=bn(T(i,1)) | T(i,4) | T(i,6);          % on the boundary
  bn(T(i,2))=bn(T(i,2)) | T(i,4) | T(i,5);
  bn(T(i,3))=bn(T(i,3)) | T(i,5) | T(i,6);
  K(T(i,1:3),T(i,1:3))=K(T(i,1:3),T(i,1:3))+Ke;     % assemble
  M(T(i,1:3),T(i,1:3))=M(T(i,1:3),T(i,1:3))+Me;
end;
b=M*feval(f,N(:,1),N(:,2));                          % right hand side and
for i=1:m,                                           % boundary conditions
  if bn(i)>0,
    b(i)=feval(g,N(i,1),N(i,2));
    K(i,:)=zeros(1,m); K(i,i)=1;
  end;
end;
u=K\b;
```

Using the MATLAB commands

```
[N,T]=NewMesh(2);
[N,T]=RefineMesh(N,T);
N=SmoothMesh(N,T);
[N,T]=RefineMesh(N,T);
N=SmoothMesh(N,T);
[N,T]=RefineMesh(N,T);
N=SmoothMesh(N,T);
engine=inline('400*exp(-100*((1.05-x).^2+(0.125-y).^2))','x','y');
friction=inline('exp(-(0.5*(x-0.1).^2+80*(y-0.05).^2))','x','y');
u=FEPoisson(engine,friction,N,T);
PlotSolution(u,N,T);
axis('equal');view([0 90]);
```

we obtain the result shown in Figure 5.9. To visualize the solution, we used the `PlotSolution` function

```
function PlotSolution(u,N,T,P);
% PLOTSOLUTION plots a solution given on a triangular mesh
%   PlotSolution(u,N,T,P); plots the solution vector u on the mesh
%   given by nodes N and triangles T also coding u using color.
%   Note that often view([0 90]) then gives a good 2d color contour plot.
%   If the fourth argument P is given, then according to P the color is
%   shifted

for i=1:size(T,1),
```

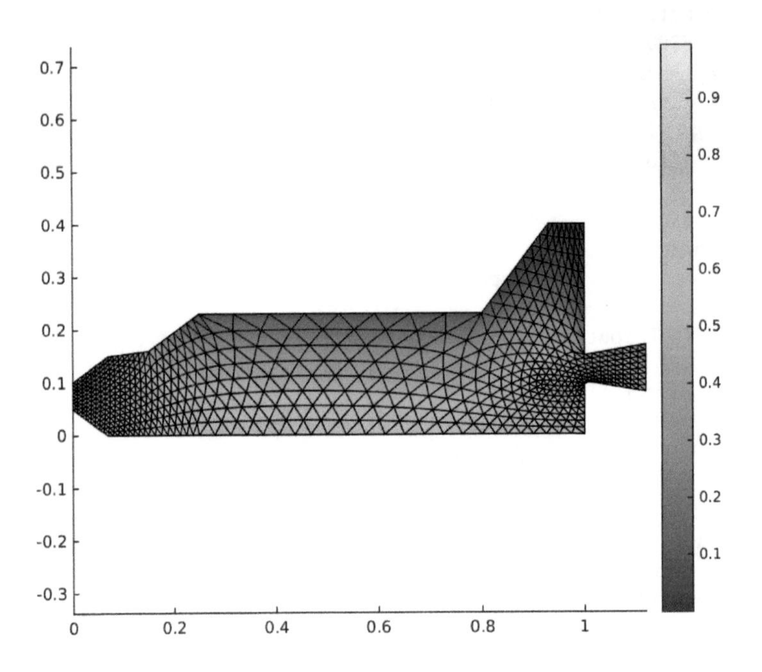

Figure 5.9. *Finite element solution of the Poisson equation on a domain representing a space shuttle.*

```
  if nargin==4 & ~isempty(P) & P(i)<=1
    patch(N(T(i,1:3),1),N(T(i,1:3),2),u(T(i,1:3)),u(T(i,1:3)),'EdgeColor','none')
  else
    patch(N(T(i,1:3),1),N(T(i,1:3),2),u(T(i,1:3)),u(T(i,1:3)));
  end
end;
colorbar;
```

Remark 5.6. *The assembly procedure we have shown in* `FEPoisson` *is quite slow when one uses a very refined mesh. A much faster assembly can be achieved by replacing the line*

```
K=sparse(m,m); M=sparse(m,m);
```

by the three lines

```
ii=zeros(n,9); jj=zeros(n,9);                % 9 entries for element matrice
Ka=zeros(n,9); Ma=zeros(n,9);
[I,J]=ndgrid(1:3,1:3); I=I(:)'; J=J(:)';
```

and the the two lines

```
K(T(i,1:3),T(i,1:3))=K(T(i,1:3),T(i,1:3))+Ke;
M(T(i,1:3),T(i,1:3))=M(T(i,1:3),T(i,1:3))+Me;
```

by the two lines

```
ii(i,:)=T(i,I); jj(i,:)=T(i,J);              % assemble vectors
Ka(i,:)=Ke(:)'; Ma(i,:)=Me(:)';              % for speed
```

and then assembling the matrices using the `sparse` *command in MATLAB by putting between* `end;` *and* `b=M*feval(f,N(:,1),N(:,2));` *the line*

```
K=sparse(ii,jj,Ka); M=sparse(ii,jj,Ma);        % fast sparse assembly
```

Further speed can be gained by working on the transposed matrices when inserting the Dirichlet conditions,

```
K=K';                      % only for speed: deleting columns is faster!
for i=1:m,
  if bn(i)>0,
    b(i)=feval(g,N(i,1),N(i,2));
    K(:,i)=0;K(i,i)=1;
  end;
end;
K=K';
```

With these modifications, the shuttle example refined seven times to get a mesh with 81920 *elements, and* 41537 *nodes can be assembled and solved in* 12 *seconds on a laptop, while the first version takes* 133 *seconds to solve the same problem. Refining once more, we arrive at* 327680 *elements and* 164993 *nodes, and the problem can be solved in* 2.5 *minutes with the fast assembly procedure, while it takes* 32 *minutes using the first version. With the fast assembly procedure, this small code becomes a quite efficient and versatile base for computing finite element approximations.*

5.8 ▪ Concluding Remarks

Finite element methods are the most general methods currently available for the approximate solution of PDEs, and they are now used in virtually any area of applications. Finite element methods have the following advantages and disadvantages, denoted by plus and minus signs, respectively:

+ The finite element method gives a systematic way to obtain discretizations for PDEs.

+ The finite element method works for arbitrary geometries and meshes.

+ The finite element method is based on the solid mathematical foundation of variational calculus.

+ Higher-order convergence can be obtained by using higher-order or spectral finite elements.

– The method is a bit more complicated to implement than a finite difference method.

There are many finite element software packages; a Wikipedia page dedicated to this is currently showing over 40, among them FreeFem++, which is especially suited for the rapid testing of finite element simulations, since the problem can be specified directly in terms of its variational formulation, and many finite elements are available.

5.9 ▪ Problems

Problem 5.1 (strong form, weak form, and minimization for Neumann boundary conditions). We consider the problem of finding $u \in H^1(a,b)$ which minimizes the functional

$$J(u) := \int_a^b \left[p(u')^2 + ru^2 - 2fu\right] dx - 2[u(b)B + u(a)A], \tag{5.43}$$

where $p \in \mathcal{C}^1([a,b])$, $p > 0$ and $r, f \in \mathcal{C}^0([a,b])$, $r > 0$ and A, B are two constants.

1. Show that the minimization problem (5.43) is equivalent to the following variational problem: Find $u \in H^1(a,b)$ such that $\forall v \in H^1(a,b)$,

$$\int_a^b [pu'v' + ruv]\, dx = \int_a^b fv dx + v(b)B + v(a)A. \tag{5.44}$$

2. Show that the variational problem (5.44) is equivalent to the strong formulation: Find $u \in \mathcal{C}^2(a,b) \cap \mathcal{C}^1([a,b])$ such that

$$\begin{cases} -(p(x)u')' + r(x)u = f(x), & \text{in } \Omega = (a,b), \\ -p(a)u'(a) = A, \\ p(b)u'(b) = B. \end{cases} \tag{5.45}$$

Problem 5.2 (advection-diffusion equation). We want to solve the advection-diffusion equation

$$\nu u_{xx} + a u_x = f, \quad \text{in } \Omega = (0,1), \tag{5.46}$$

with $\nu > 0$ and homogeneous Dirichlet boundary conditions.

1. Derive the variational form of the problem.
2. Use hat functions to define a Ritz–Galerkin approximation and compute the stiffness matrix K of the problem.
3. Compare the stiffness matrix K to the discretization matrix one obtains using centered finite differences. Explain your findings.
4. Compute the mass matrix M of the problem to get the complete linear system

$$K\mathbf{u} = M\tilde{\mathbf{f}}.$$

5. Implement the method in MATLAB and test it for different values of ν, a, and f.

Problem 5.3 (Poisson equation with Neumann boundary conditions). Consider the Poisson equation

$$\begin{cases} -u_{xx} = f & \text{on } \Omega = (0,1), \\ u_x(0) = 0, \\ u_x(1) = 0. \end{cases}$$

1. Show that this problem has a solution only if the compatibility condition $\int_0^1 f dx = 0$ is satisfied and that in this case the solution is not unique. What condition can one impose on the solution u to obtain uniqueness?

2. Derive the variational formulation of this problem, and obtain again the compatibility condition from the variational formulation.

3. Derive the minimization formulation of the problem and again from it the compatibility condition.

Problem 5.4 (stiffness and mass matrix). Consider the Helmholtz equation

$$\begin{cases} \Delta u + k^2 u = 0 & \text{in } \Omega, \\ u = g & \text{on } \partial\Omega. \end{cases} \tag{5.47}$$

1. Show that from the discretized variational formulation: Find $u \in V_h$ such that

$$-(u', v') + k^2(u, v) = (f, v) \qquad \forall v \in V_h,$$

where $V_h = span\{\varphi_1, \ldots, \varphi_n\} \in V$, one obtains the linear system

$$-K\mathbf{u} + k^2 M\mathbf{u} = \mathbf{b},$$

where K is the stiffness matrix and M is the mass matrix and $\mathbf{b}$ contains the Dirichlet boundary conditions.

2. Discuss the existence and uniqueness of a solution to the above variational problem.

3. Assume that a triangulation of the domain Ω is given. Compute on a triangular element with nodes (x_1, y_1), (x_2, y_2), (x_3, y_3) the element stiffness and mass matrices using Maple.

Problem 5.5 (simulating a chicken in a microwave). The goal of this problem is to simulate the heating of a chicken in the microwave. Instead of using the full Maxwell's equations in three dimensions, we will first introduce some simplifications that give rise to a two-dimensional Helmholtz equation, which we then solve using the finite element method. For a result, see Figure 5.10.

1. The electric and magnetic field in a microwave oven satisfy Maxwell's equation

$$\begin{aligned} \nabla \times \mathbf{E} &= -\mu \mathbf{H}_t, \\ \nabla \times \mathbf{H} &= \varepsilon \mathbf{E}_t + \sigma \mathbf{E}, \end{aligned}$$

where the vector $\mathbf{E}$ denotes the electric field and the vector $\mathbf{H}$ the magnetic field and ε is the permittivity, σ the conductivity, and μ the permeability. In air we have $\varepsilon_a = 8.85e-12$, $\sigma_a = 0$, and $\mu_a = \pi 4e-7$, whereas in chicken the values are $\varepsilon_c = 4.43e-11$, $\sigma_c = 3e-11$, and $\mu_c = \pi 4e-7$.

   (a) In a microwave, the source is time harmonic, i.e., a multiple of $e^{i\omega t}$ for $\omega = 2\pi f$, where $f = 2.45e9$ is the frequency. The solution is therefore also time harmonic,

$$\mathbf{E} = \mathbf{E}(\mathbf{x})e^{i\omega t}, \quad \mathbf{H} = \mathbf{H}(\mathbf{x})e^{i\omega t}.$$

   Show that in this case Maxwell's equation becomes

$$\begin{aligned} \nabla \times \mathbf{E} &= -i\omega\mu \mathbf{H}, \\ \nabla \times \mathbf{H} &= i\omega\tilde{\varepsilon} \mathbf{E} \end{aligned}$$

   for a complex parameter $\tilde{\varepsilon}$, which you also need to derive.

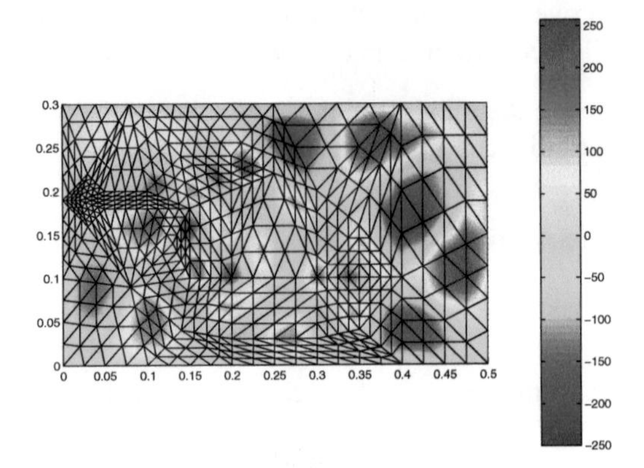

Figure 5.10. *Simulation of a chicken heating in a microwave. Note how the chicken contains hot spots; this is why microwave ovens have turntables and why more modern ovens change the frequency slightly while heating. Observe that hot spots are concentrated in a layer of about* 3 *cm from the surface and are rare in the center of the chicken. This shows that the microwave oven heats from the outside toward the inside, and the waves will penetrate further only when the chicken meat is completely dry. So the chicken is too big to be cooked in a microwave oven.*

(b) If we assume now that the variation in the z direction is small, i.e., $\partial_z \mathbf{E} = \partial_z \mathbf{H} = 0$, derive a new equation for the third component of the electric field E_3:

$$\Delta E_3 + \omega^2 \mu \tilde{\varepsilon} E_3 = 0.$$

This equation is called the Helmholtz equation.

2. Implement and test the simple mesh generator described in this chapter.

3. Implement the general finite element code explained in this chapter using the fast assembly procedure from Remark 5.6, and test your code on the square with various refinements using as boundary condition $g = x + y$ and $\omega = 0$. What result should you get? Does your computational result agree with this?

4. Simulate now the situation of the chicken cooking in the microwave. The microwave source can be modeled by the boundary condition

```
g = inline('100*(x==0.5 & 0.1<=y & 0.2>=y)','x','y');
```

You should get a result like the one shown in Figure 5.10. Is this enough resolution? Try to refine and compare. Can you explain what you observe?

Bibliography

[1] B. Andreianov, F. Boyer, and F. Hubert. Discrete duality finite volume schemes for Leray-Lions type elliptic problems on general 2D-meshes. *Numerical Methods in PDEs*, 23:145–195, 2007. (Cited on p. 84)

[2] S. C. Brenner and L. R. Scott. *The mathematical theory of finite element methods*. Texts in Applied Mathematics 15. Springer Science+Business Media, 2008. (Cited on pp. 1, 123, 125, 127)

[3] W. L. Briggs, V. E. Henson, and S. F. McCormick. *A Multigrid Tutorial*, 2nd ed., SIAM, 2000. (Cited on p. 50)

[4] C. G. Canuto, M. Hussaini, and A. Quarteroni. *Spectral Methods in Fluid Dynamics*. Springer-Verlag, 1987. (Cited on p. 1)

[5] G. Ciaramella and M. J. Gander. *Iterative Methods and Preconditioners for Systems of Linear Equations*. In preparation, 2017. (Cited on p. 50)

[6] J. Clerk Maxwell. On Faraday's lines of force. *Transactions of the Cambridge Philosophical Society*, 10:27, 1864. (Cited on p. 31)

[7] R. W. Clough. The finite element method in plane stress analysis. In *Proc. ASCE Conf. Electron. Computat.*, Pittsburgh, PA, 1960. (Cited on p. 114)

[8] R. Courant. Variational methods for the solution of problems of equilibrium and vibrations. *Bulletin of the American Mathematical Society*, 49:1–23, 1943. (Cited on p. 114)

[9] R. Courant, K. Friedrichs, and H. Lewy. Über die partiellen Differenzengleichungen der mathematischen Physik. *Mathematische Annalen*, 100(1):32–74, 1928. (Cited on pp. 46, 51, 59)

[10] R. Dautray and J.-L. Lions. *Analyse mathématique et calcul numérique pour les sciences et les techniques*. Collection du commissariat à L'Energie Atomique. Série Scientifique. Masson, 1984–1985. (Cited on p. 34)

[11] R. Dautray and J.-L. Lions. *Mathematical Analysis and Numerical Methods for Science and Technology: Volume 1. Physical Origins and Classical Methods*. Springer Science+Business Media, 2012. (Cited on p. 34)

[12] S. Delcourte, K. Domelevo, and P. Omnes. *Discrete Duality Finite Volume Method for Second Order Elliptic Problems*. Hermes Science Publishing, pp. 447–458, 2005. (Cited on p. 84)

[13] J. d'Alembert. Recherches sur la courbe que forme une corde tendue mise en vibrations. *Academie Berlin*, pp. 214–249, 1747. (Cited on pp. 28, 30)

[14] J. d'Alembert. Recherches sur les vibrations des cordes sonores. *Opuscules Matématiques*, 1:1–73, 1761. (Cited on p. 30)

[15] W. Engelhardt. Ohm's law and Maxwell's equations. In *Annales de la Fondation Louis de Broglie*, Vol. 41, pp. 39–53, 2016. (Cited on p. 32)

[16] L. C. Evans. *Partial Differential Equations.* Graduate Studies in Mathematics 19. AMS, 2002. (Cited on pp. 121, 124)

[17] R. Eymard, T. Gallouët, and R. Herbin. Finite volume methods. *Handbook of Numerical Analysis*, 7:713–1018, 2000. (Cited on pp. 1, 80, 83, 84)

[18] P. Forsyth and P. Sammon. Quadratic convergence for cell-centered grids. *Applied Numerical Mathematics*, 4(5):377–394, 1988. (Cited on p. 83)

[19] J. Fourier. *Theorie analytique de la chaleur.* Chez Firmin Didot, père et fils, 1822. (Cited on pp. 18, 19, 20, 22, 87)

[20] B. G. Galerkin. Rods and plates: Series occurring in various questions concerning the elastic equilibrium of rods and plates. *Engineers Bulletin (Vestnik Inzhenerov)*, 19:897–908, 1915. (in Russian). (Cited on pp. 89, 114, 115)

[21] M. Gander. Iterative methods for Helmholtz and Maxwell equations. *Oberwolfach Proceedings*, 2013. (Cited on p. 37)

[22] M. J. Gander and G. Wanner. From Euler, Ritz, and Galerkin to modern computing. *SIAM Review*, 54(4):627–666, 2012. (Cited on p. 114)

[23] S. Gerschgorin. Fehlerabschätzung für das differenzenverfahren zur lösung partieller differentialgleichungen. *ZAMM—Journal of Applied Mathematics and Mechanics/Zeitschrift für Angewandte Mathematik und Mechanik*, 10(4):373–382, 1930. (Cited on pp. 46, 51)

[24] D. J. Griffiths. *Introduction to Electrodynamics*, 3rd ed., Prentice Hall, 1999. (Cited on p. 32)

[25] B. Gustafsson, H.-O. Kreiss, and J. Oliger. *Time Dependent Problems and Difference Methods.* Wiley InterScience, 1996. (Cited on p. 1)

[26] E. Hairer, C. Lubich, and G. Wanner. *Geometric numerical integration: Structure-preserving algorithms for ordinary differential equations*, Vol. 31. Springer Science+Business Media, 2006. (Cited on p. 15)

[27] E. Hairer, S. Nørsett, and G. Wanner. *Solving Ordinary Differential Equations* I. *Nonstiff Problems.* Springer-Verlag, 1987. (Cited on p. 15)

[28] E. Hairer and G. Wanner. *Solving Ordinary Differential Equations* II, Vol. 1. Springer-Verlag, 1991. (Cited on p. 15)

[29] E. Hairer and G. Wanner. *Analysis by Its History.* Springer Science+Business Media, 2008. (Cited on p. 90)

[30] H. Helmholtz. Theorie der Luftschwingungen in Röhren mit offenen Enden. *Journal für reine und angewandte Mathematik*, 57:1–72, 1859. (Cited on p. 37)

[31] A. Hurwitz and R. Courant. *Vorlesungen über die allgemeine Funktionentheorie und elliptische Funktionen.* Julius Springer, 1922. (Cited on p. 114)

[32] C. Johnson. *Numerical Solution of Partial Differential Equations by the Finite Element Method.* Cambridge University Press, 1987. (Cited on p. 1)

[33] I. Kleiner. *Excursions in the History of Mathematics: The State Space Method*, Vol. 178. Springer Science+Business Media, 2012. (Cited on p. 30)

[34] H.-O. Kreiss and J. Oliger. Comparison of accurate methods for the integration of hyperbolic equations. *Tellus*, 24(3):199–215, 1972. (Cited on p. 90)

[35] C. Lanczos. Trigonometric interpolation of empirical and analytical functions. *Journal of Mathematics and Physics*, 17(1):123–199, 1938. (Cited on p. 88)

[36] L. Lapidus and G. F. Pinder. *Numerical Solution of Partial Differential Equations in Science and Engineering*. Wiley InterScience, 1999. (Cited on p. 1)

[37] E. Leonhard. Principes généraux du mouvement des fluides. *Académie Royale des Sciences et des Belles-Lettres de Berlin, Mémoires*, 11:274–315, 1757. (Cited on p. 34)

[38] R. J. Leveque. *Finite Volume Methods for Hyperbolic Problems*. Cambridge Texts in Applied Mathematics, Cambridge University Press, 2002. (Cited on p. 1)

[39] J. Liesen and Z. Strakos. *Krylov Subspace Methods: Principles and Analysis*. Oxford University Press, 2013. (Cited on p. 50)

[40] A. J. Lotka. *Elements of Physical Biology*. Williams & Wilkins, 1925. (Cited on pp. 9, 10)

[41] S. Lui. *Numerical Analysis of Partial Differential Equations*, Vol. 102. John Wiley & Sons, 2012. (Cited on p. 1)

[42] R. W. MacCormack and A. J. Paullay. Computational efficiency achieved by time splitting of finite difference operators. American Institute of Aeronautics and Astronautics, Aerospace Sciences Meeting, 10th, San Diego, Calif., Jan. 17–19, 1972, 8 p. 1972. (Cited on pp. 69, 70)

[43] P. W. McDonald. The computation of transonic flow through two-dimensional gas turbine cascades. In ASME TurboExpo: Power for Land, Sea, and Air, *ASME* 1971 *International Gas Turbine Conference and Products Show*, American Society of Mechanical Engineers, 1971, V001T01A089, doi:10.1115/71-GT-89. (Cited on pp. 69, 70)

[44] T. Narasimhan. Fourier's heat conduction equation: History, influence, and connections. *Reviews of Geophysics*, 37(1):151–172, 1999. (Cited on p. 19)

[45] I. Newton. Principia mathematica. *Newton's principia*, 1686. (Cited on p. 6)

[46] E. M. Purcell. *Electricity and Magnetism*. McGraw-Hill, 1985. (Cited on p. 32)

[47] A. Quarteroni and A. Valli. *Domain Decomposition Methods for Partial Differential Equations*. Oxford Science Publications, 1999. (Cited on p. 50)

[48] L. F. Richardson. *Weather Prediction by Numerical Process*. Cambridge: University Press, 1922. (Cited on p. 46)

[49] W. Ritz. Über eine neue Methode zur Lösung gewisser Variationsprobleme der mathematischen Physik. *Journal für die reine und angewandte Mathematik (Crelle)*, 135:1–61, 1908. (Cited on pp. 87, 114, 115)

[50] W. Ritz. Theorie der Transversalschwingungen einer quadratischen Platte mit freien Rändern. *Annalen der Physik*, 18(4):737–807, 1909. (Cited on pp. 87, 114)

[51] J. A. Ruffner. Reinterpretation of the genesis of Newton's "Law of Cooling." *Archive for History of Exact Sciences*, 2(2):138–152, 1963. (Cited on p. 17)

[52] C. Runge. Über eine methode die partielle differentialgleichung $\delta u =$ Constans numerisch zu integrieren. *Zeitschrift für Mathematik und Physik*, 56:225–232, 1908. (Cited on p. 46)

[53] Y. Saad. *Iterative Methods for Sparse Linear Systems*, 2nd ed., SIAM, 2003. (Cited on p. 50)

[54] B. Smith, P. Bjorstad, and W. Gropp. *Domain Decomposition: Parallel Multilevel Methods for Elliptic Partial Differential Equations*. Cambridge University Press, 2004. (Cited on p. 50)

[55] A. Steiner and M. Arrigoni. Die lösung gewisser räuber-beute-systeme. *Studia Biophysica*, 123(2), 1988. (Cited on p. 9)

[56] A. Steiner and M. J. Gander. Parametrische lösungen der räuber-beute-gleichungen im vergleich. *Il Volterriano*, 7:32–44, 1999. (Cited on p. 9)

[57] G. Strang and G. J. Fix. *An Analysis of the Finite Element Method*, Vol. 212. Prentice Hall, 1973. (Cited on p. 1)

[58] J. C. Strikwerda. *Finite Difference Schemes and Partial Differential Equations*. Chapman & Hall, 1989. (Cited on pp. 1, 52)

[59] V. Thomée. From finite differences to finite elements: A short history of numerical analysis of partial differential equations. *Journal of Computational and Applied Mathematics*, 128(1):1–54, 2001. (Cited on p. 46)

[60] A. Toselli and O. Widlund. *Domain Decomposition Methods—Algorithms and Theory*. Springer Series in Computational Mathematics 34. Springer, 2004. (Cited on p. 50)

[61] L. N. Trefethen. *Spectral Methods in MATLAB*. SIAM, 2000. (Cited on pp. 1, 107)

[62] N. J. Turner, R. W. Clough, H. C. Martin, and L. J. Topp. Stiffness and deflection analysis of complex structures. *Journal of Aeronautical Sciences*, 23:805–23, 1956. (Cited on p. 114)

[63] V. Volterra. Variazioni e fluttuazioni del numero d'individui in specie animali conviventi. *Memoirs of The Academy of Lincei Roma*, 2:31–113, 1926. (Cited on p. 9)

Index

An *italic* page number indicates that the term appears as part of an exercise in a problem section.

advection term, 27, 60
  centered discretization, 60
  downwind discretization, 60
  upwind discretization, 60
advection-diffusion equation, 16, 27, 37, 59, 60, 68
advection-reaction-diffusion equation, 27, 59
Alembert, J. le Rond d', 28
Ampere's law, *see* Maxwell–Ampere law
assembly
  finite element, 137, 140, 142
  finite volume, 84

backward Euler, 36
boundary condition, 16, 17, 19, 20, 30, 49, 69
  Dirichlet, *see* Dirichlet boundary condition
  homogeneous, 22, 24, 30, 42
  Neumann, *see* Neumann boundary condition
  Robin, *see* Robin boundary condition
  with finite differences, 56
  with finite volumes, 69, 71
bounded variation, 90

centered scheme, 61
Chebyshev
  points, 102
  polynomial, 103, *109*
  spectral method, 102
conservation laws, *42*
control volume, 69, 70
convection-diffusion equation, 27
  see also advection-diffusion equation
crude oil production, 10
curl, 5, 33
curl-free, 36
current density, 32

differential operator, 5, 58
  order, 58
differentiation matrix
  Chebyshev spectral, 104
  finite difference, 95
  Fourier spectral, 95, 96
diffusion, 26, 27
  equation, 19, 25, 37
  see also heat equation
Dirichlet boundary condition, 24, 46, 49
  with Chebyshev spectral, 106
  with finite differences, 49, 56
  with finite elements, 121
  with finite volumes, 71, 72
discrete cosine transform, 104, *110*
discrete maximum principle, 53, *68*
divergence, 5, 33
  theorem, 19, *42*
downwind scheme, 59

electric permittivity, 32, 37
element stiffness matrix, 137
elliptic problems, 34
essential boundary condition, 121
Euler, Leonhard, 35
  constant, *39*
  incompressible equations, 34
explicit method, 36

Faraday's law, *see* Maxwell–Faraday law
finite difference
  centered, 47, 57, 59
  convergence, 51
  downwind scheme, 59
  five-point star, 48, 49, 55
  ghost point, 57
  in time, 35
  method, 1, 45, 55
  nine-point star, 55
  one-sided, 56–59
  program, 64
  relation with finite volumes, 74
  upwind scheme, 59
finite element
  convergence, 125
  method, 1, 114
  program, 137
  shape functions, 136
finite volume
  cell centered, 72, 74, 85
  convergence, 80
  method, 1, 69
  program, 83
  relation with finite differences, 74
  vertex centered, 72, 85
forward Euler, 36
Fourier
  coefficient, 89, 94
  discrete transform, 94
  Fast Fourier Transform (FFT), 96
  law of heat flux, 17
  series, 22, 88, 93
  spectral method, 95
Fourier, Joseph, 18
FV4 scheme, 78

Galerkin approximation, 117, 125, 130
Gauss law, 33
Gauss–Green theorem, *41*
  see also divergence theorem
ghost point, 57
gradient, 5
  weak, 123
grid function, 53

hat functions, 118
heat equation, 16, 17, 19, 24, 36
  time-harmonic, 37
  see also diffusion equation
Helmholtz equation, *145*
  see also wave equation
Hooke's law, 28

initial condition, 6, 16, 17, 20, 30
initial value problem, 6
isoparametric element, 137

Laplace equation, 25, 36
  see also Poisson equation
Laplacian, 5
  discrete, 48
local truncation error, 55
Lotka, Alfred J., 9
Lotka–Volterra model, 9, *41*

magnetic permeability, 32
Maple, *38*
  curl, 5
  D (operator), 3, 4
  derivative, *see* diff (function)
  diff (function), 2, 4
  divergence, 5
  dsolve, 7
  expression, 4
  functions, 4
  gradient, 5
  Laplacian, 5
  linalg package, 5
  ODE solution, *39*
  partial derivative, 3
  sequence operator $, 2
mass matrix, 118, 136
  global, 137
MATLAB, *38*
  data fitting, 12
  delsq, 65
  fft, 96, 102, 106
  Fourier coefficients, 96
  ifft, 102
  linear system, 39
  numgrid, 62
  ODE solution, 8, *39*
  ode45, 8, 13
  plotting, 39
  sparse, 83, 143
  spy, 64
Maxwell's equations, 31, *145*
Maxwell, James C., 32
Maxwell–Ampere law, 32
Maxwell–Faraday law, 32, 36
mesh, 69, 70
  arbitrary, 69
  Cartesian, 74
  cell, 119
  dual, 76
  generator, 130, 146
  parameter, 58, 62
  point distribution, 110
  primal, 76
  rectangular, 48, 70
  refined, 142
  refinement, 83, 84, 114, 126, 134
  size, 48, 55, 59, 65, 93, 127
  smoothing, 134
  tangling, 135
  triangular, 114
  uniform, 48
  visualization, 132
minimization formulation, 119

natural boundary condition, 121
Navier, Claude-Louis, 34
  Navier–Stokes equations, 17, 33, 37
Neumann boundary condition, 24, 46
  with Chebyshev spectral, *111*
  with finite differences, 56, 67
  with finite elements, 121
  with finite volumes, 72, *85*
Newton, Isaac, 7
  law of cooling, 17, *40*
  law of motion, 6, 28, 29
  Principia Mathematica, 7

ODE, *see* ordinary differential equation
Ohm's law, 32
order
  differential operator, 58
  local truncation error, 67, 68
  multi-index, 15
  of differentiation, 15
  partial differential equation, 15
  Runge–Kutta method, 8
ordinary differential equation (ODE), 1, 6
  reduction to first order, 6
Oseen equations, 34

partial derivative, 3
partial differential equation (PDE), 1, 15
  classification, 15
  elliptic, 16, 25
  hyperbolic, 16, 30
  order, 15
  parabolic, 16, 19
  separation of variables, 20, *42*
PDE, *see* partial differential equation, 87
pendulum, 6, *39*
  analytical solution, 8
  implicit solution, 7
periodic boundary condition, 88
  compatibility condition, 89
  with finite differences, *108*
  with Fourier spectral method, 95
Poisson equation, 16, 25, 36, 46, 51, 56
population dynamics, *see* Lotka–Volterra model
predator-prey interaction, *see* Lotka–Volterra model

reaction equation, *see* advection-reaction-diffusion equation
reaction term, 27
reference triangle, 137
Ritz approximation, 119, 130
Robin boundary condition, 57
  centered, 57
  one-sided, 57
  with finite differences, 57
  with finite volumes, 73
Runge–Kutta method, 8

semidiscretization, 35
shape functions, 136, 137
  local, 139
shifted Laplace equation, 37
sparse
  linear system, 50, 71
  matrix, 50

spectral
  Chebyshev-based method, 102, *110*
    program, 106
  convergence, 90, 93, 100
  Fourier-based method, 88, 95
    convergence, 98
    program, 101
  method, 1, 87
steady state solutions, 36
stiffness matrix, 117, 120, 121, 136
  element, 137
  global, 137
Stokes, George G., 34
  Navier–Stokes equations, 17, 33
  Stokes equation, 37
strong form, 115, 120, 129
structured
  linear system, 71
  matrix, 50

Taylor series, 38, 47, 62, 66, 76, 79, 81, 88
test function, 120, 122, 127, 129
  space, 120, 121, 125
time-harmonic, 37
total variation, 90
TPFA, *see* two-point flux approximation
trial function, 121
  space, 120
truncation error, 48
  estimate, 52
two-point flux approximation (TPFA), 71, 78

upwind scheme, 59, 61

variational form, *see* weak form
Volterra, Vito, 9
  Lotka–Volterra model, 9, *41*
Voronoi cells, 70

wave equation, 16, 28
  time-harmonic, 37
  see also Helmholtz equation
weak form, 115, 117, 120, 130